Asia-Pacific Security

ASIA-PACIFIC SECURITY

An Introduction

Joanne Wallis &
Andrew Carr, Editors

Georgetown University Press   *Washington, DC*

Library of Congress Cataloging-in-Publication Data

Names: Wallis, Joanne, editor. | Carr, Andrew, editor.
Title: Asia-Pacific security : an introduction / Joanne Wallis and
    Andrew Carr, editors.
Description: Washington, DC : Georgetown University Press, 2016. |
    Includes bibliographical references and index.
Identifiers: LCCN 2015047169 (print) | LCCN 2015048575 (ebook) | ISBN
    9781626163447 (hc : alk. paper) | ISBN 9781626163454 (pb : alk. pa-
    per) | ISBN 9781626163461 (eb) | ISBN 9781626163461
Subjects: LCSH: Security, International—Asia—Textbooks. | Security,
    International—Pacific Area—Textbooks. | Asia—Foreign relations—
    21st century—Textbooks. | Pacific Area—Foreign relations—21st
    century—Textbooks.
Classification: LCC JZ6009.A75 A44 2016 (print) | LCC JZ6009.A75 (ebook)
    | DDC 355/.03305—dc23 LC record available at http://lccn.loc.gov
    /2015047169.

17 16   9 8 7 6 5 4 3 2 First printing

Printed in the United States of America

Cover design by Pam Pease

Cover images clockwise from top left:
Hong Kong Skyline © Daumiu / Shutterstock
US Defense Secretary Ash Carter meets Japanese Prime Minister Shinzō
    Abe on April 8, 2015. Photo courtesy of US Department of Defense
Sydney Harbour Bridge © Taras Vyshnya / Shutterstock
Soldiers © US Department of Defense, MC1 Chad McNeeley
Globe © Tito Onz / Shutterstock

Contents

Illustrations

Preface

This book is written primarily as a textbook for undergraduate and graduate university students. It is intended to be a core text for courses related to security in the Asia-Pacific or international relations of the region. It has a strong theoretical component and therefore will also be useful as a supplementary text in more general courses related to security studies or the Asia-Pacific region. However, the expertise of its authors means that it also constitutes an important contribution to the growing academic literature on security studies and security in the region.

Despite the sophistication of its content, this book is deliberately written to be accessible to the student reader; thus, it avoids jargon and the chapters incorporate key security studies themes that arise across the region. Each chapter opens with a reader's summary, important terms appear in boldface the first time they are used in each chapter and are defined in the glossary, and each chapter concludes with a summary of key points, a list of questions, and a reading guide.

Acknowledgments

This book grew out of the security studies teaching program at the Strategic and Defense Studies Centre of the Australian National University (ANU). The ANU is an inspiring and supportive research and teaching environment, and we are both proud to work there. The excitement of our students about studying security in the Asia-Pacific and their enthusiasm for accessing a resource that provides a comprehensive introduction and overview of the region motivated us to edit this book.

We gratefully acknowledge the support of our colleagues at the Strategic and Defense Studies Centre—particularly its head, Brendan Taylor, for his generous intellectual and financial support. The Centre sits within the ANU's College of Asia and the Pacific, and we thank the College Education Committee for its grant toward the production of this book. That grant allowed us to employ Anna Samson and Josh Wyndham-Kidd to provide research assistance to finalize the book, and we thank them for their careful work. This book also features a number of maps of the Asia-Pacific region, for which we thank the ANU's CartoGIS.

To contribute the chapters to this volume, we were fortunate to be able to bring together leading scholars who work at the forefront of debates about security in the Asia-Pacific. The book would not have been possible without these scholars, who generously devoted their time to producing their excellent contributions.

We have also been fortunate to have Donald Jacobs as our editor at Georgetown University Press. Working with Don has demonstrated the value of an engaged and thorough editor and a comprehensive review process. This book was considerably improved from its first draft as a result of Don's work and the reviewers' comments, for which we are also grateful.

Finally, Joanne thanks her husband, Ross Mulcahy, whose love, understanding, and support were critical to the production of this book. Andrew thanks his wife, Katina Curtis, for her love and patience with his unorthodox use of grammar.

Abbreviations

ADMM+	ASEAN Defense Ministers' Meeting Plus
APC	Asia-Pacific Community
APEC	Asia-Pacific Economic Cooperation (forum)
ARF	ASEAN Regional Forum
ASEAN	Association of Southeast Asian Nations
ASEAN+3	ASEAN plus China, Japan, and South Korea
CCP	Chinese Communist Party
CSCAP	Council for Security Cooperation in the Asia-Pacific
EAS	East Asia Summit
EEZ	Exclusive Economic Zone
FAO	Food and Agriculture Organization of the United Nations
GDP	gross domestic product
IMF	International Monetary Fund
JI	Jemaah Islamiyah (Islamist group)
NATO	North Atlantic Treaty Organization
NSA	National Security Agency (United States)
PLA	People's Liberation Army (China)
PPP	purchasing power parity
R2P	responsibility to protect
RAMSI	Regional Assistance Mission to Solomon Islands
SLOC	sea lanes of communication
TPP	Trans-Pacific Partnership (trade agreement)
UNCLOS	United Nations Convention on the Law of the Sea
UNSC	United Nations Security Council

An Introduction to Asia-Pacific Security

Andrew Carr and Joanne Wallis

Reader's Guide

This chapter introduces the security environment of the contemporary Asia-Pacific. This chapter also introduces the main theories that are used in this book as it investigates the region and its security challenges. These theories, drawn from security studies, include "traditional" approaches (realism and liberalism) and nontraditional ones (constructivism, the Copenhagen School, critical security studies, and human security). Each of these theories is analyzed by asking three key questions: *Whose security is the focus? What threats to security are identified? How can security can be achieved?* These theories are then applied to contemporary issues in the Asia-Pacific to illustrate their relevance to studies of security in the region.

Introduction

This book provides an introduction to the security environment of the Asia-Pacific in the early twenty-first century. The importance of this region was demonstrated in 2011, when the world's most powerful country, the United States, announced that it would "pivot" or rebalance its military, economic, and political focus toward the Asia-Pacific. The region is also home to emerging global powers, China and India; to increasingly activist regional powers, such as Japan; and to security flashpoints that could have global ramifications, including the South China Sea, the Korean Peninsula, and the Taiwan Strait. This book discusses the key players, provides an in-depth examination of the traditional and nontraditional security challenges, and outlines some of the proposed solutions to improve the security and stability of the region.

This chapter begins by describing why policymakers and scholars are increasingly looking at the Asia-Pacific as the center of international affairs. It then highlights the main security theories that guide this book as it investigates the region and its security challenges. These theories provide a framework for analyzing regional events and for predicting what might happen in the future.

There is currently no agreed-on definition of the "Asia-Pacific" region; it potentially covers a majority of the world's population, stretching from India and Pakistan to the United States (map I.1). However, agreeing to a definition is important, for it determines which states are able to become members of the region's institutions and which are seen as legitimate voices in shaping how the region operates. Even the definition of "Asia" is controversial; some scholars believe that it has "always been a Western, or European, concept, not an idea born or fostered in the region itself," because Asia "makes no particular ethnic or racial sense."[1] Others disagree, arguing that Asia was not "simply the invention of Westerners, [but] was also robustly imagined from within Asia" by Asian leaders.[2] There are also debates over which label should be used to describe the region. China and Malaysia tend to favor a narrower conception of "East Asia." Japan prefers a slightly broader panregional conception of "Asia," whereas the United States favors the more expansive notion of the "Asia-Pacific." Recently, the rise of India has drawn significant international attention toward the Indian Ocean, leading states such as Australia to suggest the term "Indo-Pacific."[3]

This debate over terminology reflects the fact that, as power shifts, so do regional identities and geographic boundaries. It also reflects the direction and attention of state power and diplomacy. Some states such as Canada continue to take an interest in Asia, but may not in the future. Likewise, much of South America is removed from the debates and issues of Asia, but its Pacific-bound countries such as Colombia, Chile, and Peru could one day seek a larger role. Central Asian states can also be considered part of the Asia-Pacific, yet have little interaction with most states in the region. These countries do not feature prominently in this book, but it is not unreasonable to expect a new edition far in the future revising based on nearby states choosing to focus more or less on this part of the world. For the purposes of clarity, this book adopts the more common term "Asia-Pacific." It adopts a definition that focuses on littoral Asia and key Pacific states, and identifies four subregions (maps I.2–I.4):

Northeast Asia: China, Japan, the Korean Peninsula, and part of Russia;
Southeast Asia: the ten member states of the Association of Southeast Asian Nations (ASEAN);
South Asia: India, Pakistan, Bangladesh, Sri Lanka, Nepal, Bhutan, the Maldives, and the Indian Ocean region; and
The broader Pacific Ocean region: the United States, Canada, Australia, New Zealand, and the Pacific islands.

Why Is the Asia-Pacific Region Important?

In November 2011 the Obama administration announced that the United States would "pivot" toward the Asia-Pacific, which marked its formal recognition that

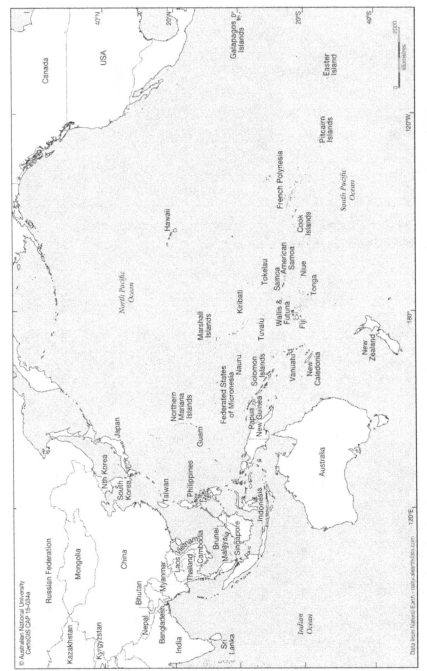

Map I.1 The Asia-Pacific. © The Australian National University, College of Asia and the Pacific, CartoGIS.

Map I.2 Northeast Asia. © The Australian National University, College of Asia and the Pacific, CartoGIS.

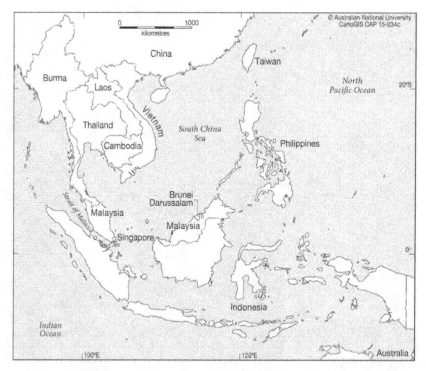

Map I.3 Southeast Asia. © The Australian National University, College of Asia and the Pacific, CartoGIS.

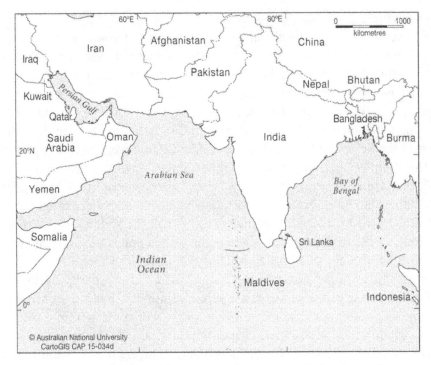

Map I.4 South Asia and the Surrounding States. © The Australian National University, College of Asia and the Pacific, CartoGIS

the region had entered a new era in which the existing security order is evolving. In a speech to the Australian Parliament, President Obama said:

> As the world's fastest-growing region—and home to more than half the global economy—the Asia Pacific is critical . . . with most of the world's nuclear powers and some half of humanity, Asia will largely define whether the century ahead will be marked by conflict or cooperation, needless suffering or human progress.[4]

As the president noted, the Asia-Pacific has witnessed significant economic growth during the last thirty years. States such as Japan, South Korea, Singapore, China, and India have progressed toward developed-world conditions of life expectancy, education, infrastructure, and national wealth. In practical terms, this meant that during the first six years of the twenty-first century (which is sometimes described as the "Asian Century"), about 1 million people were brought out of poverty each week in Northeast and Southeast Asia.[5]

This growth offers both opportunities and challenges. Economic growth opens up the potential for greater trade and new social, cultural, and political interactions. However, states with growing economies can also afford to spend more on their militaries. China has been increasing its defense spending by 140 percent since 2000, which affects the global distribution of military power.[6] Accordingly, many European states have followed the United States in seeking to improve their position in the Asia-Pacific, and states on its fringes, such as Russia and Australia, have refocused on the region. All want to share in the region's growing prosperity, but they are also concerned that the past half century of relative peace in the region might be coming to an end. As attention shifts to the region, people in the Americas, Europe, and the Middle East are worried that they might lose the attention and support of the United States. To assuage these concerns, the Obama administration renamed its "pivot" a "rebalance," to lessen the perception that the United States was withdrawing its interests from other parts of the world. However, as crises continue to escalate in the Middle East, increasing questions are being asked about whether the United States is committed to the rebalance, or whether its attention still lies primarily elsewhere.

In order to sustain the record levels of growth and prosperity in the Asia-Pacific, new security arrangements are needed. States will need to avoid conflict over precious resources; prevent nontraditional security risks from distracting or destabilizing them; and ensure that rising states are able to obtain the voice and recognition they seek (and often deserve) without creating tension with the established powers of the region. These and many other problems will be at the heart of the research in the field of security studies for decades to come. This book provides tools to analyze these issues.

What Is Security Studies?

The Asia-Pacific is experiencing changes to its economics, politics, demographics, culture, and social organization. The security order in the region is the foundation upon which these other changes take place. The discipline of security studies attempts to explain how this order operates. It is a subdiscipline of international relations, which in turn is a subdiscipline of political science. The specific focus on security studies emerged in the aftermath of World War II as an effort to assist Western governments to understand the emerging nuclear era. Today it has expanded beyond a narrow focus on war to cover a broad range of traditional and nontraditional challenges, although it still retains a focus on academic insights into vital policy challenges.

What Is "Security"?

The basic definition of security is "freedom from threats to core values for both individuals and groups."[7] Accordingly, something is secure when it has the capacity to defend itself against sources of danger at a reasonable cost. However, there are three disagreements over the definition of "security," which concern:

- The **referent object** of security (i.e., *whose security?*);
- The scope of security (i.e., *security from what?*); and
- The approach to security (i.e., *how to secure?*).

In the following paragraphs, each of these questions is used to test the main theories of security studies discussed in this chapter. Where key terms are used for the first time, they are highlighted in **bold** and an explanation is provided in the glossary (which appears at the end of the book).

Why Is Theory Useful for Security Studies?

It is important to use theoretical approaches in security studies, because they provide the analytical frameworks that help us to make sense of the changing world. These frameworks enable us to decide which facts matter and how they relate to each other. They also provide us with common understandings of key terms and ideas, which allows for comparisons between different cases. Theoretical approaches are particularly useful for scholars of Asia-Pacific security studies, as they provide tools to analyze the structural changes under way in the region.

Traditional Approaches to Security

There are two broad theories that represent traditional approaches to security. The first is **realism**, which has been by far the most influential.

In answer to the question *whose security?* realists argue that states are the primary referent object of security. They assume that states are unitary actors that act according to the same self-interested impulses, regardless of how they organize their political, economic, social, or religious systems. The core concern for realism is the survival of states. Realists define survival in political terms as encompassing "the integrity of the nation's territory, of its political institutions, and of its culture."[8] Only if states are secure can the scourge of war be prevented.

In answer to the question *security from what?* realists argue that security is freedom from the threat of armed attack and freedom from fear of an armed attack. They hold that the international system is defined by **anarchy**, by which they mean that there is no global law enforcement authority to manage international conflict, enforce agreements, and guarantee the survival of states. Realists view **power** as the defining feature of the anarchic international system. Without an international authority to protect them, states need to rely on their own power to achieve their international goals. Power is generally understood as the resources available to a state for building military forces, including wealth, population, and technological sophistication.

Realists differ in their answer to the question *how to secure?* Classical realists, who claim a history of scholarship that is more than 2,000 years old, assert that human nature is flawed and the desire for power is the ultimate cause of conflict and war. As the international system is defined by anarchy, classical realists say that statesmen should act rationally to assure the survival of their state by maximizing their state's power and seeking to expand its territory.

Neorealism emerged in the 1970s as an attempt to provide a more sophisticated way to explain how states interact and an attempt to bring order to anarchy. Neorealists see the anarchic structure of the international system as more important than human nature in causing conflict between states. Because states are uncertain of each other's intentions, they must resort to **self-help** to guarantee their survival. That is, states must take care of their own security by strengthening their military power and by forming alliances with other states to balance against threats.[9] Neorealists see material factors (especially the distribution of power and structure of the international system) as more important than ideational factors (i.e., states' internal characteristics, the nature of political systems, ideas, identities, interests, and norms) in determining a state's actions.

Moreover, neorealists hold that all power gains are relative (i.e., for every gain one state makes, another loses, which is known as a **zero-sum** game). Because states engage in self-help, this leads to a "**security dilemma**," whereby a state's desire for security forces it "to acquire more and more power in order to escape the impact of the power of others. This, in turn, renders the others more insecure and compels them to prepare for the worst."[10] This means that instability, conflict, and war remain a constant possibility.

Neorealists argue that it is possible to achieve stability within the international system in three ways. First, a **hegemony** may emerge, under which a single state maintains order by superior economic and military resources, dominates the system, defines collective goals and rules, and enforces them. However, when the hegemon declines, or new challengers arise, instability, conflict, and war are common. Second, a **strategic condominium** or **concert of powers** can develop when two or more great powers manage international affairs and accommodate their competing interests on the basis of common goals, values, and interests. Finally a **balance of power** can operate, when two or more great powers with similar capabilities balance each other in a "bipolar" or "multipolar" order and thereby prevent conflict.

Under each of these models, neorealists say that less powerful states have two options. First, they can attempt balancing, whereby they either form alliances with other states, to draw on those other states' resources, or they seek to increase their own power to balance the existing great powers. Second, they can engage in bandwagoning, whereby they join one of the strong powers (or the only strong power).

Arguing in favor of the United States' hegemony in the Asia-Pacific, Aaron Friedberg argues that the United States remains the world's leading economic and military power by a significant margin, although he acknowledges that in the future China is likely to challenge it for power and influence in the region and globally.[11] However, this raises the question of whether the United States has ever been a hegemon in the region, as states like China and India have not necessarily acquiesced to the United States' dominance of the region. In contrast to Friedberg, Michael Leifer and Henry Kissinger have argued that a balance of power, most likely between the United States and China, could emerge in the region.[12] A weakness of the balance of power is that it is highly fragile as intense competition remains a possibility and states remain highly attentive to shifts in the constellation of power. This means that a balance of power almost inevitably breaks down and often ends up requiring major war to restore equilibrium. Alternatively, Amitav Acharya and Hugh White have proposed that a strategic condominium or concert of powers could emerge, whereby the region's great powers could have a common interest in cooperating to manage the region's security challenges.[13] However, a concert of powers requires a degree of policy and ideological agreement that may be difficult to achieve between the United States and China, or between China and Japan, for the reasons explained in chapters 1, 2, and 3.

Offensive realists are skeptical about the potential for creating a security order because they believe that anarchy means that states can never be certain about each other's intentions and must therefore compete with each other.[14] As a result, they argue that states attempt to maximize their power relative to other states, because only the strongest states can guarantee their survival. In contrast, defensive realists argue that anarchy does not necessarily create conflict between states, with variations of cooperation and reassurance being possible.[15] As such, they argue that, in some circumstances, to diffuse the security dilemma a state's best option is cooperation or restraint so that its capabilities are more equally balanced with those of other states.

The potential for states to cooperate to ensure their security is an idea strongly taken up by the second traditional theoretical approach to security: **liberalism**. Liberals are optimistic about the prospects for cooperation among international actors and the prospects of a peaceful world. In answer to the question *whose security?* liberals agree that the state is the primary referent object of security, however, they also pay attention to other actors, such as international institutions, nongovernmental organizations, multinational corporations, and domestic actors.

In answer to the question *security from what?* liberals agree that states need to be secured from the threat or fear of armed attack. However, though liberals recognize international anarchy, they are optimistic that its effects can be mitigated so that international security is achieved. Of particular importance, liberals do not see power acquisition as a zero-sum game, so they do not believe a security dilemma is inevitable. In contrast to realism, which sees all states as "black boxes," liberalism argues that all states are different (i.e., some are democracies, and some are authoritarian), so ideational factors are as important as material ones.

Like realists, liberals differ in their answer to the question *how to secure?* Economically focused liberals argue that a security order can be created through economic **interdependence** via international trade, investment, and people movements. They argue that economic interdependence reduces incentives for international conflict as the costs of the use of force rise while the benefits decline. States will therefore prefer "to trade than to invade."[16] Scholars such as Richard Rosecrance argue that economic interdependence between the United States and China, between Japan and China, and between Taiwan and China is likely to mitigate the risk of insecurity in the region, as the costs of conflict are now too great.[17] However, economic interdependence failed to prevent the European powers from spiraling into world war in 1914, and a persuasive case can be made that the Asia-Pacific today exhibits a number of strong similarities to Europe on the eve of World War I. For example, **nationalism** is currently on the rise, particularly in Northeast Asia, and many regional states are engaging in significant military **modernization** programs. Economic interdependence is, in many respects, driving the economic growth that is allowing them to buy these weapons.

Neoliberals build on the structural ideas of neorealists but see their ideas as supporting a liberal worldview. They argue that the negative effects of anarchy can be mitigated by states cooperating to form international **institutions**, which "can provide information, reduce transaction costs, make commitments more credible, establish focal points for coordination and, in general, facilitate the operation of reciprocity."[18] Neoliberals argue that institutions perform three roles: They are constitutive (i.e., they help define states' interests); they can regulate states' behavior through rules and conventions; and they can facilitate peaceful changes to the structure of the international system. Neoliberals thus argue that a relatively high degree of international governance and order is possible. Liberal scholars point to the growing institutionalization of the Asia-Pacific as a positive sign that a stable security order may emerge. Although few multilateral institutions existed before the 1990s, there are now dozens of forums operating in the region, including the East Asia Summit, the Asia-Pacific Economic Cooperation forum, the Shanghai Cooperation Organization, the ASEAN Regional Forum, and the Shangri-La Dialogue. John Ikenberry argues that a rising China's behavior can be shaped or "socialized" by integrating it as a member of the United States–led liberal order and its supporting institutions.[19] Similarly, Evelyn Goh argues that ASEAN has succeeded in "enmeshing" great powers into ASEAN-centered structures such as the ASEAN Regional Forum, through which it has exposed them to regional norms.[20] However, regional institutions have been slow-moving and unable to respond effectively to major crises (e.g., the 1989–2013 Korean nuclear crisis, the 1995–96 Taiwan Strait crisis, the 1997–98 Asian financial crisis, and the 2004 Indian Ocean tsunami).

It is also questionable whether institutions create order or whether they ultimately reflect the prevailing order. As discussed in chapter 11, there is a concern in the Asia-Pacific that the great powers are not so much cooperating through the same institutions as they are competing via the institutions within which they have the most influence—the United States through Asia-Pacific Economic Cooperation and its own ad hoc arrangements, such as the **Proliferation Security Initiative**; China through **ASEAN+3** and the Shanghai Cooperation Organization; and Japan through the ASEAN Regional Forum and the East Asia Summit.

The ideal institution, according to neoliberals, is a **security community**, which exists when states cooperate in order to achieve collective security on the basis of agreed-on principles. The idea is that when states integrate their interests, a preference for peaceful conflict resolution take precedence over state power. This prevents aggression against any member state through the credible threat of collective action, including military force. Amitav Acharya argues that a security community is emerging in Southeast Asia, as none of the original ASEAN members have fought one another since the organization was first formed in 1967.[21] However, border clashes remain a relatively frequent occurrence, and Southeast Asian states have been buying up conventional weaponry, including sophisticated fighter jets and submarines, which raises the question of whether there is the risk of an **arms race** in the region, as is considered in chapter 6. The Southeast Asian example also raises questions regarding whether the security community concept could apply across the entire Asia-Pacific. The region may not have the required sense of collective identity, the similarity of domestic political systems, or states with the willingness and ability to renounce the use of force to settle disputes.[22] It may also have different security trends operating in different subregions. For these reasons, as explored in the book's conclusion, there may not be a neat single solution to the region's security challenges.

Liberals argue that if states have liberal-democratic political systems, and there is international cooperation and interdependence, then long-term peace is possible. This is known as the "**democratic peace thesis**," and is based on the observation that democracies do not go to war against each other. This thesis rests on three arguments. First, the "constitutional caution" argument holds that electorates pay the price of war and are therefore likely to constrain their governments from going to war through democratic checks and balances. Second, the "global economic interdependence" argument holds that, due to international trade and investment, peace becomes too profitable for states to endanger by going to war. And third, the "mutual identification" argument holds that democratic states are more likely to give each other the benefit of the doubt during episodes of diplomatic crisis than they are when engaging with nondemocratic states.

If the liberal arguments are right, it might be possible for a disparate group of states, in a condition of anarchy, to build a **security order**; that is, a situation where "interaction among states is not arbitrary but conducted in a systematic manner on the basis of certain rules."[23] The most important elements of a security order are rules, because they make for a predictable and stable environment in which states can work together to pursue their goals and in which disputes between states can be adjudicated in a peaceful manner, without resorting to violence.

The rules that govern a security order may be explicit and codified, or they may be informal and be inferred from consistent state practice. For an order to be present, these rules must be acknowledged and states must behave in accordance with these rules in the knowledge that there will be a cost if they violate them. International law, as set out in the Charter of the United Nations, is an example of rules that are acknowledged by most states and that most states obey most of the time. For example, virtually all states acknowledge that they will be in violation of international law if they breech the sovereignty of another state by invading it. This suggests that, despite the inherent insecurity that realists identify, with proper institutions, structures, and willingness, it may be possible to create a peaceful global society. In the words of Hedley Bull, a global society would involve a situation where "a group of states, conscious of certain common interests and common values, . . . conceive themselves to be bound by a common set of rules in their relations with one another," and thus help to overcome the problem of anarchy and insecurity.[24] This type of society does not yet exist, but liberals can point to many significant strides toward this ideal that have been made in the last fifty years.

Nontraditional Approaches to Security

Realism and liberalism are "positivist" theories, given that they purport to describe how the world is. In contrast, nontraditional approaches are "post-positivist," because they question prevailing social and power relationships and institutions and argue that no theory is politically neutral. Nontraditional approaches have emerged as security studies has begun to question the assumptions made by the traditional approaches, deepening its understanding of what should be secured, and broadening the range of threats and dangers it identifies. Moreover, though realism and liberalism are "problem-solving" theories, which accept the world they have inherited and seek to make it work, nontraditional approaches are generally "normative," which means that they seek to describe how the world could—and should—be.

The most influential nontraditional approach is **constructivism**, which emerged after the end of the Cold War, out of a recognition that the traditional approaches had not been able to predict, or adequately comprehend, why the Cold War finished and how the international security order had been transformed. In contrast to realists, constructivists say that ideational factors are as important as material power. In fact, they say that material power only acquires meaning through discourse and shared understandings—a process known as social construction. Typifying this, Alexander Wendt famously stated that "a gun in the hands of a friend is a different thing from one in the hands of an enemy."[25]

Constructivists say that identities matter because they create interests, and these interests tell us how actors behave and the goals that they pursue. Therefore, variations in states' identities affect their interests and policies. Unlike realists or liberals, constructivists argue that state interests are socially constructed and thus can change over time. Whereas realists and liberals argue that the international system is inherently based on anarchy, constructivists argue that shared ideas, beliefs, and values also have structural characteristics, and that they exert a powerful influence on social and political action within the international system. Constructivism therefore provides us with a way to understand states' interests, states' behavior, and the international system by trying to explain the actions, beliefs, and interests of states and other actors.

Constructivists vary in their answer to the question *whose security?* While they broadly agree that the referent object of security is states, there are three main approaches to studying them. First, systemic constructivists (e.g., Wendt) focus on the way that states interact in the international system. Second, "unit-level" constructivists (e.g., Peter Katzenstein) focus on states but are concerned with domestic politics, as well as the relationship between domestic social and legal norms and the identities and security interests of states. Unit-level constructivists try to explain why two states adopt different internal and external national security policies. Third, holistic constructivists (e.g., John Ruggie and Friedrich Kratochwil) try to bridge the divide between unit-level constructivism and systemic constructivism. As a result, they try to explain how state identities and interests are constituted and how this leads to shifts in the international system.

In answer to the question *security from what?* constructivists argue that security interests and threats are socially constructed as a result of states' identities, interests, and interactions. Constructivism can help to explain how states define threats to their national security, ideas of security order, and threats to that order. Thus, constructivists see nothing "inevitable" about great power conflict, while cautioning that factors such as culture and history can often play destructive roles by setting states on a path toward conflict.

Similarly, in answer to the question *how to secure?* constructivists say that both the international security order and actors within it (including states) are socially constructed based not just on material power but also on ideational factors. Variations in these factors affect states' interests, policies, and behavior—not just the pursuit of power. So, in contrast to realists, constructivists say that, if we live in an anarchical international system, it is because states have come to believe that is how the world is, and their actions correspond to that reading of an anarchic world. This led Wendt to argue that "anarchy is what states make of it."[26] Therefore, constructivism offers hope of changing the international system, as it holds that interests are not fixed and instead new norms can enter collective understandings and recreate the international security order. This is important for security studies, as it suggests that we can move beyond the logic of anarchy.

The fact that so many questions relating to identity, ideas, and norms remain unresolved suggests that constructivism may be useful for analyzing security in the diverse and complex Asia-Pacific region. In particular, whereas the traditional approaches were formulated primarily to describe the behavior of states in Europe and North America, constructivism allows us to ask how and why Asia-Pacific states might behave differently. For example, constructivism helps us to explain why there are tensions between states such as China and Japan, by highlighting factors such as history and identity, thereby providing a more sophisticated analysis than a mere calculation of their relative power. It is also integral to understanding the way particular security orders from hegemonies to security communities are developed and reinforced. However, it is harder for policymakers to draw clear policy responses from constructivist analyses than more traditional approaches to studying security.

Like constructivism, the **Copenhagen School** seeks to reconceptualize the notion of security. In answer to the question *whose security?* it argues that referent objects are "things that are seen to be existentially threatened and that have a legitimate claim to survival."[27] Thus, referent objects can be the state but also other "human collectivities," national economies, ideologies, collective identities, species, or habitats.

In answer to the question *security from what?* the Copenhagen School agrees with traditional theories that security is about survival, but it identifies five sectors of security: military, political, economic, societal, and environmental. The most important aspect of the Copenhagen School is its concept of "**securitization.**" This concept says that we should treat security as a "speech act," that is, what determines whether something is a security issue is whether people describe it as a security issue and if those who listen accept that description. The securitization of an issue occurs when actors declare that a concern is a security issue and that a referent object is threatened. Once a matter has been securitized, it is prioritized above "normal politics," and "extraordinary means" are necessary to address the problem. The Copenhagen School answers the question *how to secure?* in a similar way as realists, for it maintains the security-survival logic of the traditional approaches. However, because it identifies a wider range of sectors that can be threatened, it extends the tools of achieving security beyond military power.

What is missing from many traditional analyses is an acknowledgment that security threats may not only involve military means but may also involve new issues such as transnational crime, terrorism, disease, and environmental challenges, as chapters 8 and 9 outline. Many of these new issues have been securitized and have therefore become state security issues. The Copenhagen School is useful here as it gives us tools to explore the reasons why certain new issues become securitized and others do not, and whether they actually post any greater threat than other issues.

Like constructivism and the Copenhagen School, **critical security studies** theorists are interested in the role of ideas. They argue that the social world is produced in and through ideas that make it meaningful, which are themselves necessarily social. Therefore, states and other actors are social constructs, which are constituted through political practices. As a result, critical security studies does not agree with realism that the international system is anarchical, because like constructivism it says that it is socially constructed. However, critical security studies does not denote a coherent approach to security. Instead, it represents a desire to move beyond the traditional approaches to security. Because it holds that all knowledge is socially constructed, critical security studies allows us to see that, though the traditional realist and liberal theories claim to be neutral, in fact they are also social constructs.

In answer to the question *whose security?* critical security studies critically examines the traditional focus on states in order to identify hidden factors. Critical security studies allows us to ask whether we should only be concerned with the state and its security. In general, critical security studies conceives security "comprehensively, embracing theories and practices at multiple levels of society, from the individual to the whole human species."[28] By looking at individuals and the communities where they live as the referent objects of security, critical security studies provides scope to take seriously the ideas, norms, and values that constitute the communities (including states) that are to be secured.

Critical security studies answers the question *security from what?* by asking what security might mean from different perspectives. In particular, though states might be primarily threatened by the use and control of material power, people and their collectivities can be threatened in all sorts of ways, including by their own states. Consequently, critical security studies also answers the question *how to secure?* by recognizing that the international security order is a social construct, and it therefore questions what it should look like from different perspectives. As a result, critical security studies provides possibilities for reconstructing perceptions of the international system so that states no longer see it as being based on anarchy. If states can move beyond seeing the international system as anarchical, they may also be able to stop engaging in self-help, which may transform the security dilemma. Critical security studies is also helpful because it forces us to think through the reasons why states, groups, and people act in the way that they do, which can in turn help us to identify more effective responses.

Building on the ideas generated by critical security studies, the **human security** approach emerged after the Cold War, when the end of US-Soviet rivalry saw a decline in conventional military threats, while other threats, such as those to human well-being and the environment, came to the fore. There was also a recognition that the nature of war had changed from "old wars" fought between states by armed forces in uniform to "new wars" fought by networks of state and nonstate actors, usually without uniforms, and with violence often directed against civilians.[29] The concept of human security helps us to deal with the security challenges posed by new wars, as it allows us to recognize that "new wars" frequently involve other threats, such as terrorism, transnational crime, the spread of disease, vulnerability to natural disasters, poverty, and homelessness.

In answer to the question *whose security?* the human security approach, explored at length in chapter 12, argues that the referent object of security should be people, or individuals.[30] While traditional theories focus on the security of states, the human security approach recognizes that part of the reason we focus on states is their role protecting their citizens. Yet states can, and do, abuse their power by violating human rights and threatening the security of their citizens. Moreover, in many conflicts, violence is not perpetrated by states but instead by nonstate actors such as guerrilla organizations and militias, insurgents, secessionist and terrorist groups, and transnational crime organizations. Some states are not necessarily legitimate and cohesive actors and may be unable or unwilling to protect their citizens against such violence. In an increasingly globalized world, states are also becoming less able to provide physical security, material welfare, and a habitable environment. The human security approach helps call attention to these factors and provides tools with which to study their implications.

Human security approaches differ in their answer to the question *security from what?* The narrow school focuses on "freedom from fear" from the threat of violence to people by the state or any other organized political actors. In contrast, the broad school argues that freedom from fear also encompasses "freedom from want" arising from underdevelopment and threats to other human freedoms. Therefore, for the broad school human security means, "first, safety from such chronic threats as hunger, disease and repression. And second, it means protection from sudden and hurtful disruptions in the patterns of daily life."[31] This division means that the human security approach differs in its answer to the question *how to secure?* The narrow school advocates measures to manage the threat of political violence, including humanitarian interventions, peacekeeping missions, **counterinsurgency**, and the promotion of democracy. The broad school adds to this by also advocating measures aimed at improving people's material sufficiency so that their basic needs are met. The straightforward but important message of human security is that states should not always be the primary referent of security when examining the Asia-Pacific. The human security approach highlights the fact that poverty, underdevelopment, violence within states (e.g., ethnic conflicts or secessionist movements), terrorism, transnational crime, pandemics, and environmental change may have a more significant impact on many people living in the region than state-level security threats such as great power rivalry and interstate flashpoints.

Conclusion

The security environment of the Asia-Pacific appears to be in a state of flux, with important changes to the security order of the region being generated by the rise of other powers, such as China and India and the United States' pivot to the region. The main theories that guide security studies provide tools to analyze and seek to understand these changes. Although no single theory offers a comprehensive analysis of recent events in the region, scholars and policymakers draw from multiple security theories when attempting to understand and shape the region.

Structure of the Book

This book adopts a three-part structure. Part I highlights the *actors* that shape the Asia-Pacific: the United States (chapter 1), China (chapter 2), India and Japan (chapter 3), the middle powers (chapter 4), and the small states (chapter 5). Part II examines the current and emerging *security challenges* that these states face, including arms races (chapter 6); maritime security (chapter 7); terrorism and insurgencies (chapter 8); nontraditional security challenges, such as climate change, disease, and migration (chapter 9); and changing technology, especially cybersecurity and space security (chapter 10). Part III concludes by examining the *proposed solutions* that will enable the actors to overcome the security challenges they face and preserve the region's prosperity. Two ideas for keeping the region peaceful that are examined in depth are multilateralism and security communities (chapter 11), and human security (chapter 12). The conclusion analyzes several models that have been applied to assess the future security order of the region.

Key Points

- The Asia-Pacific has become central to international affairs since the United States' pivot to the region and the rise of the emerging global powers China and India.
- Traditional and nontraditional security studies theories can be applied to events in the region in order to analyze their meaning, to make predictions about the future, and potentially to offer solutions to ensure the security of the Asia Pacific.
- Each security studies theory has strengths and weaknesses, which suggests that care needs to be taken when applying them to analyze the Asia-Pacific.
- Scholars are attempting to use security studies theories to understand how the Asia-Pacific is developing. Some are optimistic; some are pessimistic. But all agree that the region will be vital for global security and prosperity.

Questions

1. Why is the security of the Asia-Pacific becoming increasingly important to global security?

2. What are the strengths and weaknesses of the traditional security studies theories?

3. What alternatives does the application of nontraditional security studies theories offer when analyzing the security of the Asia Pacific?

4. Which security studies theory (or theories) provides the most convincing account of current events in the region?

5. How might a security studies theory (or theories) be used to identify solutions that will ensure the security of the region in the future?

Guide to Further Reading

Alagappa, Muttiah, ed. *Asian Security Practice: Material and Ideational Influences*. Stanford, CA: Stanford University Press, 1998.
  This book consists of chapters that explore the relevance of nontraditional security studies theories to sixteen states in the Asia-Pacific. It focuses on the security thinking and behavior of elites in the region, in order to determine how they behave.

Booth, Ken. *Critical Security Studies and World Politics*. Boulder, CO: Lynne Rienner, 2005.
  This book outlines the key elements of critical security studies and illustrates how they can be applied to contemporary security issues.

Miller, Benjamin. "The Concept of Security: Should It Be Redefined?" *Journal of Strategic Studies* 24, no. 2 (2001): 13–42. http://dx.doi.org/10.1080/01402390108565553.
  This article undertakes a critical analysis of existing definitions of "security" and proposes alternative definitions that may be of more utility to future analysis.

Snyder, Jack. "One World, Rival Theories." *Foreign Policy* 145, no. 145 (2004): 52–62. http://dx.doi.org/10.2307/4152944.
  This article builds on the summary provided by Stephen M. Walt (see the next work listed here) to include updated analysis of the key security studies theories.

Walt, Stephen M. "International Relations: One World, Many Theories." *Foreign Policy* 110 (1998): 29–46. http://dx.doi.org/10.2307/1149275.
  This article summarizes the key security studies theories in a clear and coherent manner.

Notes

1. Bill Emmott, *Rivals: How the Power Struggle between China, India, and Japan Will Shape Our Next Decade* (Orlando: Harcourt, 2008), 34.

2. Amitav Acharya, "The Idea of Asia," *Asia Policy* 9 (2010): 33.

3. Australian Department of Defence, *Defence White Paper 2013* (Canberra: Commonwealth of Australia, 2013).

4. Barack Obama, "Speech to Parliament," *The Australian*, November 17, 2011.

5. Indermit Gill and Homi Kharas, *An East Asian Renaissance: Ideas for Economic Growth* (Washington, DC: World Bank, 2007).

6. Australian Department of Defence, *Defence White Paper 2013*, 9.

7. John Baylis, "International and Global Security in the Post–Cold War Era," in *The Globalization of World Politics*, ed. John Baylis, Steve Smith, and Patricia Owens (Oxford: Oxford University Press, 2001), 300.

8. Hans J. Morgenthau, "Another Great Debate: The National Interest of the United States," *American Political Science Review* 46 (1952): 973.

9. Stephen M. Walt, *The Origins of Alliances* (Ithaca, NY: Cornell University Press, 1987), 21.

10. John H. Herz, "Idealist Internationalism and the Security Dilemma," *World Politics* 2 (1950): 157.

11. Aaron Friedberg, *A Contest for Supremacy: China, America, and the Struggle for Mastery in Asia* (New York: W. W. Norton, 2011).

12. Michael Leifer, Kin Wah Chin, and Leo Suryadinata, *Michael Leifer: Selected Works on Southeast Asia* (Singapore: Institute of Southeast Asian Studies, 2005); Henry Kissinger, *On China* (New York: Penguin Books, 2011).

13. Amitav Acharya, "A Concert of Asia?" *Survival* 41 (1999): 84–101; Hugh White, *The China Choice: Why America Should Share Power* (Melbourne: Black Inc., 2012).

14. John J. Mearsheimer, *The Tragedy of Great Power Politics* (New York: W. W Norton, 2001), 2.

15. Jeffrey W. Taliaferro, "Security Seeking under Anarchy: Defensive Realism Revisited," *International Security* 25 (2000): 130.

16. Uri Bar-Joseph, *Israel's National Security towards the Twenty-First Century* (London: Frank Cass, 2001), 32.

17. Richard Rosencrance, *The Rise of the Trading State* (New York: Basic Books, 1986).

18. Robert O. Keohane and Lisa L. Martin, "The Promise of Institutionalist Theory," *International Security* 20 (1995): 42.

19. G. John Ikenberry, *Liberal Leviathan* (Princeton, NJ: Princeton University Press, 2012), 6.

20. Evelyn Goh, "Southeast Asian Perspectives on the China Challenge," *Journal of Strategic Studies* 30 (2007): 810.

21. Amitav Acharya, *Constructing a Security Community in Southeast Asia* (London: Routledge, 2009).

22. Christopher B. Roberts, *ASEAN Regionalism: Cooperation, Values, and Institutionalization* (Milton Park, UK: Routledge), 10.

23. Muthiah Alagappa, *Asian Security Order: Instrumental and Normative Features* (Stanford, CA: Stanford University Press), 39.

24. Hedley Bull, *The Anarchical Society: A Study of Order in World Politics* (New York: Columbia University Press, 2002), 13.

25. Alexander Wendt, "Identity and Structural Change in International Politics," in *The Return of Culture and Identity in IR Theory*, ed. Yusef Lapid and Friedrich Kratochwil (Boulder, CO: Lynne Rienner, 1996), 50.

26. Alexander Wendt, "Anarchy Is What States Make of It: The Social Construction of Power Politics," *International Organization* 46 (1992): 395.

27. Barry Buzan, Ole Waever, and Jaap De Wilde, *Security: A New Framework for Analysis* (Boulder, CO: Lynne Rienner, 1998), 36.

28. Ken Booth, *Critical Security Studies and World Politics* (Boulder, CO: Lynne Rienner, 2005), 16.

29. Mary Kaldor, *New and Old Wars* (Stanford, CA: Stanford University Press, 2007).

30. United Nations, *Human Development Report: New Dimensions of Human Security* (New York: United Nations, 1994).

31. Ibid., 23.

Part I

The Changing Asia-Pacific Security Order

1 Can the United States Share Power in the Asia-Pacific?

Brad Glosserman

Reader's Guide

The twentieth century is often described as the "American Century." For most of it, the United States was the world's largest economic and military power, and the post–World War II order in the Asia-Pacific was largely an American creation. In the twenty-first century, however, questions have been raised about the possible decline of US power, the rise of challenger states, and whether the United States has the will and desire to sustain the order it developed. While there may be questions about whether the United States is "the indispensable power," it does remain "the indispensable partner," especially as the security landscape evolves. This chapter explores what is meant by the term "great power" and argues that the United States clearly fits this label. It then outlines the reasons behind Washington's "pivot" to Asia in 2011 and the implications for the United States and the region. Finally, the chapter explores the relationship between the United States and the People's Republic of China (hereafter, China)—especially from the US perspective—to shed light on perhaps the primary security challenge the region faces in the new century.

Introduction

Few in the US public, and none who walk its corridors of power, question the great power credentials of the United States. Madeleine Albright, secretary of state under President Bill Clinton, considered the United States to be the "indispensable power," a sentiment that grew out of her experience during the Cold War and watching Europe flail and fail while trying to deal with the bloody disintegration of Yugoslavia.[1] In recent years, however, it has become clear that great power status "ain't what it used to be." Even the United States, the sole remaining superpower, has discovered that it is increasingly unable to dictate outcomes around the world. Some blame an increasingly partisan political system or economic difficulties at home. In fact, however, twenty-first-century problems and challenges are more complex, and power and capabilities are more diffused. As the wars in Afghanistan and Iraq made clear, the United States can blow things up, but rebuilding enduring structures of peace and ensuring stability are such huge challenges that even a superpower must look for help from coalition partners.

This is the new reality of power in the twenty-first century, and one that increasingly dominates US thinking about foreign policy. In this world, a new type of leadership is required. The most apt formulation of this new reality comes from Jane Harmon, a longtime student of foreign policy who served eight terms as a member of the US House of Representatives and now heads the Woodrow Wilson International Center for Scholars. She considers the United States the "indispensable partner," a compelling and important tweaking of Albright's dictum.[2] The Obama administration's "rebalance to Asia" is the official articulation of this new type of leadership and a framework for implementing it. This chapter asks "what is a great power?" and whether the United States qualifies as such, two easily answered questions. More significantly, it probes the meaning and relevance of great power status in the twenty-first century, which invites an examination of how power works today and the changing US role in the Asia-Pacific region as it adapts to this new world. Finally, it explores the US-China relationship, identifies its major components and currents, and outlines a framework for understanding this critical bilateral relationship.

What Is a "Great Power"?

A great power is a state that is able—and is recognized as able—to exert influence around the world. This requires both the possession of interests that demand its attention and protection and the means to assert these interests. Traditionally, great powers were determined by crude measures—size of population, as well as economic and military strength. An effective political system was an implicit feature of a government able to exert its will, coupled with access to and control of the resources needed to fuel the economy and sustain the military. The historian A. J. P. Taylor provided a simple definition: "The test of a great power is the test of strength for war."[3]

The management of far-flung interests requires a capable diplomatic corps as well, and this requirement has taken on additional importance as the international system has become more complex; the number of states has been growing, institutions are proliferating, and economic and information systems are diversifying as globalization proceeds. International institutions and conferences can recognize and validate great power status. The **Congress of Vienna** explicitly identified the great powers of the nineteenth century; the possession of nuclear weapons was a symbol of great power status in the second half of the twentieth century (although the significance of that marker has diminished in recent years), as is a permanent seat on the United Nations Security Council. This reasoning and the evidence it uses reflect the realist mindset, which was explained in the preceding chapter.

The evolution of the international system—the changes just noted, as well as the growing acceptance of a normative structure rooted in individual rights and democratic values (leavened by the sovereignty of states)—has yielded new thinking about power. Traditional notions of hard power ultimately focus on an ability to compel another state to act in a particular way. Yet in an increasingly democratized world, the use of force—except in very particular circumstances— is losing legitimacy. Sheer power is no longer accepted as an acceptable rationale for the resolution of international disputes. The exercise of force to resolve such questions must be authorized, and hence legitimated, by some generally accepted international authority. Persuasion—the real thing of daily politics, not *The Godfather's* famous "offer you can't refuse"—has become a critical component of the great power foreign policy tool kit. The capacity to attract or co-opt rather than coerce or compel is what the Harvard University professor Joseph S. Nye Jr. calls "soft power."[4] Considerable confusion surrounds this concept, and it remains the subject of great dispute and debate. It is hard to measure and difficult to isolate. Nevertheless, the ability to win other governments over, to convince one state that what another state wants is also good for it, is a new prerequisite for great power status.

Is the United States Still a Great Power?

By every metric, the United States remains a great power. It is the world's leading economy, with a GDP estimated at $17.4 trillion in 2014, about 25 percent of global wealth (table 1.1). Even as the US share of global riches decreases, it remains twice as large as its leading competitor, and though the United States may yield the number one position to China, its GDP per capita is orders of magnitude larger. The US population is the third largest in the world, and, unlike the populations of many developed countries (and China), it continues to grow—a critical element of its continuing economic vitality. The US military, the most advanced and capable in human history, dwarfs all competitors; at $682 billion in 2013, its budget makes up 38 percent of global military spending and totals more than those of the next fourteen nations combined. This budget is being reduced, but the hyperbole surrounding impending cuts is much exaggerated; even after the planned reductions, the US military will still be larger than those of the next ten nations combined.

While smaller than its peak size in the Cold War, the US nuclear arsenal remains capable of destroying the world many times over, and it backstops a deterrent that spans the globe. Moreover, its great power status is validated by institutional arrangements; the United States has a permanent seat on the United

Table 1.1 The United States: Key Facts

Form of government	Democratic
GDP	$17,418.9 billion (2014)
GDP (purchasing power parity)	$17,418.9 billion (2014)
Population	319.0 million (2014)
Military spending[a]	$609.9 billion (2014)
Military personnel[b]	1,433,150 (2013)
Aircraft carriers[c]	19
Treaty allies in Asia	Australia, Japan, the Philippines, South Korea, Thailand

[a]Stockholm International Peace Research Institute (SIPRI), *SIPRI Military Expenditure Database* (Stockholm: SIPRI, 2015), http://sipri.org/research/armaments/milex/milex_database.
[b]World Bank, *Armed Forces Personnel, Total* (Washington, DC: World Bank, 2015), http://data.world bank.org/indicator/MS.MIL.TOTL.P1.
[c]John Pike, *World Wide Aircraft Carriers* (Alexandria, VA: Global Security, 2014), http://globalsecurity .org/military/world/carriers.htm.
Source: Unless otherwise specified above, data are from the Australian Department of Foreign Affairs and Trade, *United States Country Fact Sheet* (Canberra: Commonwealth of Australia, 2014), http://dfat.gov.au/trade/resources/Documents/usa.pdf.

Nations Security Council, is a member of the Group of 7 and Group of 20 economic organizations, gets to select the president of the World Bank, and is represented in every major international institution. Plainly, realist and neorealist conceptions of power validate the application of the "great power" label to the United States, even if some, such as Aaron Friedberg, believe that its power is diminishing.[5]

Equally important are US reserves of soft power. The United States is considered to be a leading proponent of Western values and an ideological structure that has enormous attraction around the world. The United States promotes democracy, the dignity of the individual, and freedom in both political and economic spheres of life.[6] This freedom spurs the creativity and energy that provide the foundation for its economic success. And though the US political system has descended into acrimony and periodic paralysis in recent years, its legitimacy is still unquestioned; US political processes represent the will of the people, another source of strength and attraction. Finally, US power and leadership are validated by the very successes of the current order, even though it is sometimes seen as a threat to America's leading position in the world. Washington was instrumental in the creation of a global economic order that enriches all nations and that promotes prosperity globally rather than in just one country. The promotion of the common good and the provision of public goods are key elements of the US claim to great power status as conceived by liberals and constructivists.

A Changing US Role in the Asia-Pacific Region

The United States is a Pacific power. The fiftieth US state, Hawaii, is located in the middle of the Pacific Ocean; the forty-ninth, Alaska, was once a part of the Russian Far East. The United States has been deeply engaged in Pacific affairs for more than one hundred and fifty years. American fingerprints are on many critical moments in Asian history: the opening of Japan by Commodore Perry (1852); the Portsmouth Treaty that ended the war between Japan and Russia (1905); the Taft-Katsura agreement that ratified Japan's presence in Korea in exchange for US annexation of the Philippines (1905); the Open Door Policy in China (1899–1949); and the division of Korea after World War II—as well as its role in the Pacific War (1941–45), the Korean War (1950–53), and the Vietnam War (1955–75). Today, the US presence occurs in five treaty alliances (with Australia, Japan, the Philippines, South Korea, and Thailand) known as the San Francisco System. This involves tens of billions of dollars in investment, hundreds of billions of dollars in trade, and tens of thousands of US citizens visiting, working, or studying in the region on any given day (table 1.2).

Throughout the Cold War, the most visible component of this US presence was the country's forward-based military forces. Largely underdeveloped, Asia was considered a secondary theater to Europe. The chief US concern was that communist-backed insurgents would establish a stronghold in the region and expand that ideology as "the dominoes fell." US alliances provided the sinews of engagement and the hard power to turn back the communist tide. In the post–Cold War world, successive US administrations struggled to articulate a coherent rationale for a continuing military presence. That effort was made considerably easier by the persistence of Cold War flashpoints—the divisions of Korea and China—as well as Asia's economic development, which has begun to transform the global balance of power and US thinking about the region.

All US presidential administrations produce a National Security Strategy when they take office.[7] Unlike that of his predecessors, President Barack Obama's strategy focused on the work that needed to be done at home as the United States struggled to recover from two devastating and drawn-out wars and the global financial crisis of 2008–9. Critical to this effort was the rejuvenation of the economy. This is the strategic and intellectual framework that produced the "pivot"—later renamed the "rebalance"—to the Asia-Pacific. The rebalance is an attempt to tap the energy of Asia, the world's most dynamic region, to promote US economic growth.[8] Thomas Donilon, US national security adviser at the end of the first Obama term, argued that:

> economically, it's impossible to overstate Asia's importance to the global economy and to our own. Asia accounts for about a quarter of global GDP at market exchange rates and is expected to grow by nearly 30 percent in 2015.

Table 1.2 US Security Treaties in Asia

Year Signed (or in Operation)	Treaty Name	Signatory Countries	Type of Treaty
Continuing treaty obligations			
1951	Mutual Defense Treaty between the United States and the Republic of the Philippines	United States and the Philippines	Bilateral
1951	Australia, New Zealand, United States Security Treaty (ANZUS)	United States, Australia, and New Zealand[a]	Trilateral
1951 (revised 1960)	Treaty of mutual cooperation and security between the United States and Japan	United States and Japan	Bilateral
1953	Mutual Defence Treaty between the United States and the Republic of Korea	United States and Republic of Korea	Bilateral
US Treaties no longer in operation			
1954–77	Southeast Asia Collective Defense Treaty, or Manila Pact (SEATO)	United States, Australia, France, New Zealand, the Philippines, Pakistan, Thailand, United Kingdom	Multilateral
1954–80	Mutual Defense Treaty between the United States and the Republic of China	United States and Taiwan	Bilateral

[a]Membership suspended in 1985 by the United States.

The region is estimated to account for nearly 50 percent of all global growth outside the United States through 2017. The region accounts for 25 percent of US goods and services exports and 30 percent of our goods and services imports. An estimated 2.4 million Americans now have jobs supported by exports to Asia, and this number is growing.[9]

Most analysis of the rebalance has focused on its geographic dimensions: the recognition that the world is becoming tripolar, with the Asia-Pacific emerging to join North America and Europe as the main centers of economic power. The rebalance acknowledges the shift in the global center of gravity toward the Pacific. Less studied, but perhaps even more important, is the reordering of priorities within the US foreign policy tool kit. In an article that laid out the thinking behind the rebalance, Secretary of State Hillary Clinton put "forward-deployed **diplomacy**," meaning a proactive engagement with the region, first on her list of the rebalance's notable elements. It was followed by more robust economic engagement, and only finally did she turn to the military.[10] It is ironic then that much of the discussion of the rebalance has focused on its military components, in particular the rotation of US marines through Darwin, Australia, the home porting of Littoral Combat Ships in Singapore, and the decision to port 60 percent of the US Navy fleet in the Asia-Pacific.

This reordering of priorities responded to new and emerging realities in Asia. Threats ranging from terrorism and climate change to pandemic diseases and piracy made plain that Cold War thinking, focused on the threat of great power war, could no longer dominate security planning; state-based threats persisted, but they were being overtaken by different challenges.[11] The budget constraints that virtually all governments face, in combination with increasingly expensive military equipment, further erode the ability of individual nation-states to act effectively by themselves. Fortunately, however, the new threats affected all nations, and a diffusion of capabilities meant that the bar for cooperation had been lowered. Moreover, the diverse nature of the threats required a range of responses; the most powerful military in human history is of less relevance and effectiveness when the foe is rising sea levels or new strains of the flu.

In this environment, the nature of international engagement changes. There is a premium put on cooperation and partnership, a context that empowers smaller nations. As a result, even great powers must now listen to the demands of other governments and respond to their needs. In many ways, the rebalance is an attempt to do just this. While Asian nations have relied on the US military presence and continue to depend on it for regional stability and security, they have long sought a more diffused form of engagement. They desire a larger US business presence and the investment that comes with it, and they have urged Washington to join and participate in regional political institutions such as the ASEAN Regional Forum and the East Asia Summit. To be sure, Asian thinking is being driven by many considerations, not least of which is the realization that deeper US engagement gives regional governments more cards to play as they engage other powers, such as China and Japan. But the bottom line is that the changing power realities of Asia demand a new form of leadership. Through the rebalance, the United States is attempting to adjust to this new environment and demonstrate a new style of engagement and leadership—one that meets Washington's needs while better responding to the concerns of its allies and partners. Some observers question the United States' ability to sustain its commitment to the rebalance in the wake of Russia's aggression against Ukraine, the overall deterioration of that relationship, and the emergence of the Islamic State threat in Syria and Iraq. However, the Obama administration has reaffirmed the centrality of the rebalance to US policy in the 2015 National Security Strategy as well as every relevant speech and testimony by US officials.

The US-China Relationship

Central to the United States' engagement with Asia is its relationship with China. The two countries had a tempestuous relationship during the first half of the Cold War. The Communist Party's seizure of power in 1949 sparked upheaval in the US foreign policy bureaucracy as scapegoats were sought in an attempt to identify "who lost China?" The enmity intensified when Chinese forces intervened on behalf of North Korea during the Korean War and helped fight the United States–led United Nations **coalition** to a bloody stalemate. The continual threats to Taiwan, where the Kuomintang government had fled in 1949, and the demands for the reunification of China sparked fears of a direct confrontation between Washington and Beijing. The hostility between the United States and China diminished after the Sino-Soviet split in the early 1960s, and by the late 1970s, until the collapse of the USSR, Washington and Beijing were pursuing a unique form of cooperation that aimed at countering and containing Moscow. Since the death of Chinese Communist Party (**CCP**) leader Mao Zedong in 1976 and the adoption of the reform policies by Deng Xiaoping in 1978, the United States has been instrumental in helping China develop its economy. The United States has opened its doors to China's best and brightest to help train and educate business professionals, bureaucrats, and policymakers. American companies have invested hundreds of billions of dollars in China, from which they have profited, but they have also helped to advance a policy that aimed to lift China out of poverty. The two countries' economies are deeply intertwined: Two-way trade reached $562 billion in 2013; China is the largest holder of US Treasury bills, with holdings in excess of $1.3 trillion at the end of 2013; and US-majority-owned companies in China were reckoned to have employed 1.4 million workers in China in 2011.[12] The two governments engage in robust dialogue and cooperation across all levels of the bureaucracy; Chinese officials claim that there are more than ninety bilateral dialogues and consultative mechanisms.[13] Most significantly, the two countries share an interest in a peaceful and stable Asia-Pacific region, and this is the case despite—or perhaps because of—the uncertainties created by US backing for the government and people of Taiwan. Given the totality of these circumstances— hundreds of thousands of Chinese students in the United States, tens of billions of dollars in US investment in China, and hundreds of billions of dollars in trade between the two countries—the charge that the United States has attempted to contain China makes little sense. At every instance, US officials repeat their commitment to a positive, constructive, and forward-looking relationship with China, and that the authoritative statements note that the United States "welcomes the rise of a stable, peaceful, and prosperous China."[14]

The relationship remains fraught, however. Despite repeated statements by both governments of a commitment to a positive relationship, there is concern that policy in both Washington and Beijing is essentially a zero-sum game, with the two sides competing for relative gains.[15] The territorial conflicts between China and Japan in the East China Sea over the Senkaku/Diaoyu Islands and between China and five other claimants in the South China Sea are seen as proof of an aggressive intent by Beijing. Chinese success in those efforts to rewrite the regional status quo is widely thought to imperil not only the interests of the rival claimants but also that of the United States. Even at times when the two countries' national interests overlap, their priorities frequently do not. Minxin Pei, one of the United States' foremost China watchers, summarized the situation by noting that the United States and China:

> share no overriding security interests or political values, and their conceptions of world order fundamentally clash. Whereas Beijing looks forward to a post-American, multipolar world, Washington is trying to preserve the liberal order it leads even as its relative power wanes. Meanwhile, numerous issues in East Asia, such as tensions over Taiwan and disputes between Beijing and Tokyo, are causing US and Chinese interests to collide more directly.[16]

Policymakers in both capitals have concluded, however, that cooperation is to be preferred to conflict. This makes sense. The United States and China are the world's two largest economies, possessors of significant nuclear arsenals as well as permanent seats on the United Nations Security Council, regional hegemons, and the core of two "civilizations" that speak for billions of people. While China is nominally communist, it poses no real ideological threat (as the USSR did). As then–national security adviser Donilon pointed out in 2012:

> there are few diplomatic and economic challenges that can be addressed in the world without having China at the table. . . . Our consistent policy has been to seek to balance these elements in a way that increases the quality and quantity of our cooperation with China, as well as our ability to compete. At the same time, we seek to manage disagreements and competition in a healthy, not disruptive, manner.[17]

China is, however, emerging as a great power. In 2007 the US economy was $14.48 trillion, four times larger than that of China ($3.494 trillion), and five years later it was only twice as large (the US economy was $16.6 trillion, while China's had grown to $8.229 trillion); the International Monetary Fund estimates that the size of the Chinese economy will surpass that of the United States between 2020 and 2025. Historically, this type of great power transition has been messy. Not only does Beijing face a regional and global order that it played almost no role in creating but China also has historical grievances and scores to settle. It has territorial conflicts with many of its maritime neighbors, and the dispute with Japan in particular remains ever fresh because of the Japanese invasion of China in the 1930s. The United States has alliances with several countries that have disputes with China (including Japan and the Philippines), and there are genuine fears that the two great powers could get dragged into a conflict as a result of actions by one of those allies.

To counter this possibility, policymakers in both governments have proposed formulas to stabilize the bilateral relationship. Among the most notable was the call by then–deputy secretary of state Robert Zoellick in 2005 for China to act as a "responsible stakeholder" in the international system. Zoellick, then number two in the US State Department, repeated that:

> the United States welcomes a confident, peaceful, and prosperous China, one that appreciates that its growth and development depends on constructive connections with the rest of the world. Indeed, we hope to intensify work with a China that not only adjusts to the international rules developed over the last century, but also joins us and others to address the challenges of the new century.[18]

Partnership, however, depended on China renouncing its hitherto passive role in international affairs and taking on a more constructive and active role. For Zoellick, Beijing had "to become a responsible stakeholder. As a responsible stakeholder, China would be more than just a member—it would work with us to sustain the international system that has enabled its success." The Chinese response to Zoellick's call was tepid. On one hand, they complained that they were being called upon to support a system that they had little role, if any, in creating. On the other hand, they bristled at the notion that another country would pass judgment on when Chinese behavior was "responsible." In both cases, China was being held to a standard that it had neither created nor accepted. The "responsible stakeholder" concept withered with the passing of the George W. Bush administration.

Still, Chinese leaders worried about the potential for conflict in power transitions, a theory seemingly validated by rising tensions between Beijing and its maritime neighbors. The Chinese government suggested that China and the United States should embrace a "new type of major country relations" that would act as a buffer against the frictions inherent in the upcoming power transition. Chinese analysts explained that this concept meant that the two governments would embrace "no conflict, no confrontation, mutual respect, and win–win cooperation" with "the goal of such a relationship [being] to strive for the two parties' peaceful coexistence, seeking common ground while reserving differences, and engaging in benign interaction and common progress."[19] The Chinese had hoped that the shirtsleeves Sunnylands Summit in June 2013 between Obama and Chinese president Xi Jinping would validate the idea and make it the conceptual framework for their bilateral relations. The United States was receptive to a tool that would blunt the sharp edges of the relationship, but endorsement of "the new type of major country relations" has not yet been forthcoming. US objections center on the lack of clarity surrounding its actual content and the concern that it appears intended to ratify a Chinese sphere of influence in East Asia. Most Chinese commentary regarding details of this concept speaks only of US action to accommodate China; there is no indication that Beijing should do anything to assuage the United States.

There is another factor behind the call for "a new type of major country relations": the desire to ensure that the US rebalance to Asia does not target China. Conversations with Chinese officials and experts underscore their fear that the rebalance is designed to check the expansion of Beijing's influence. China plays a key role in the US rebalance—but not as a target for US containment policies, as some assert. Rather, US officials reiterate at every opportunity the desire to forge "stable and productive relations" with China. The relationship with Beijing is routinely identified as one of the five pillars of the "rebalance," along with strengthening alliances, building new partnerships with emerging powers, empowering regional institutions, and helping to build a regional economic architecture to sustain shared prosperity.

This language, like all the United States' claims that it is seeking a win–win relationship with China, is viewed with suspicion in China. In fact, neither Washington nor Beijing has much faith in the other's good intentions. Both view the other with skepticism, if not suspicion; China fears that the United States seeks to limit its rise and limit its influence, though Americans worry that Beijing's ultimate objective is to dominate the region and to exclude the United States from the Western Pacific. Simply put, despite the years of cooperation and interaction, there is little trust in the relationship. And this mistrust is compounded by the usual work of militaries, which is preparing for contingencies—that is, fighting wars. In some cases—such as a crisis in Taiwan, the South China Sea, or the Korean Peninsula—US and Chinese national interests may well conflict, and prudent military planning ("hedging," in the parlance of international relations) thus looks a lot like hostile action. The two countries increasingly clash over issues. While the relationship is not defined by conflict—nor is that likely to happen—it is increasingly competitive. The most important question for the future is whether engagement will prevail over rivalry.

How to Manage the Relationship?

Beijing's call to establish "a new type of major country relations" is an attempt to ensure that cooperation wins out. Despite US resistance to this specific phraseology, it does have historical antecedents. It was the basis of the Cold War order. More recently, there have been proposals to establish a "duopoly" between the United States and China. The idea of a Group of Two—"a partnership of equals"—was floated a few years ago and gained some traction among theorists; the institutionalization of a bilateral relationship that is inherently "superior" to others would—in theory—dampen tensions by providing regular processes of interaction and channels of communication. The concept quickly lost steam, however. Though China was flattered to be recognized as a great power, many Chinese policymakers preferred not to shoulder the responsibilities that came with being part of the "two." The Americans realized that the relationship was too complex and too varied, that China was too reluctant, and that regional allies were too concerned about being slighted for the concept to work. Regional governments welcomed a process that would smooth over wrinkles in the China-US relationship, but in the very next breath they worried that decisions important to their national interests would be made without their input. As the saying goes, "The grass gets trampled when the elephants fight; the grass also gets trampled when they make love." Perhaps most important, Asian governments bristled at the idea of being forced to choose between Beijing and Washington. They have no desire to see a new bamboo curtain descend through their region. Thus they urged the two governments to find common cause and work together to ensure regional peace and stability.

At the same time, however, and somewhat confusingly, many of those same governments do worry about a newly assertive China. While they accept the geographical and geopolitical reality of China's rise, they fear that Beijing will single-mindedly pursue its own interests, and do so at their expense. Thus they seek to draw other large powers—such as the United States, Australia, and New Zealand, India, and even the European Union—into the region to increase diplomatic options and maximize leverage in negotiations. For the most part, this results in ad hoc arrangements born of the particular issue at hand. In some cases, however, this yields formal, ongoing relationships that are designed to thwart a more aggressive posture by Beijing. This hedging against Chinese misbehavior takes the form of the classic realist approach to international security threats: balancing.

When they "balance," allies, friends, and partners align themselves against China to check Beijing's freedom to maneuver and ensure that it does not become a threat to the regional status quo. While balancing is a defensive strategy, it can quickly become adversarial as relationships harden. Alliances are the classical manifestation of balancing, which is born of conventional calculations of power. Leaders assess a foreign state's interests, along with its relative strength, weigh them against their own state's attributes, and then join with the state (or group of states) whose interests most closely align with their own and can offset those of states whose interests are inimical. In short, states compensate for weakness by balancing. It should be obvious that balancing involves conventional definitions of power. In a world where there is less utility in raw power, balancing has diminished appeal. After all, even though balancing is intended to protect one state against another's hostile intentions, aligning with another state, especially one that is stronger, typically requires some sacrifice of **autonomy**. The views of the stronger ally tend to prevail.

Weaker states do not need to balance. Indeed, David Kang believes that the reflexive resort to balancing as the operational concept in international relations reflects Western parochialism. Surveying the region's history, he sees little evidence of alliance formation and balancing against China, yet considerable stability nevertheless. Instead, he argues that the traditional international order in Asia is a hierarchical system, dominated by China, in which each state had its assigned place. Historically, he suggests, China was too large, too central, and too advanced when compared with other regional states for them to balance against. In addition, China exported a set of norms and expectations that were shared around the region, as well as a model of bureaucratic governance that other states were keen to adopt (these last two factors look a lot like soft power). By accepting China's role as leading power, along with its own particular place in the order, a neighboring country won diplomatic recognition from Beijing (which could be lucrative; tribute did not always run to the center) and could enjoy the advantages of trade.[20]

While he differentiates between the two, Kang's notion of regional hierarchy approximates the better-known and more widely accepted idea of "bandwagoning." When they bandwagon, weaker states align with the dominant power to claim relative gains that they would not have either standing alone or by balancing against it. In other words, small and middle powers seek the protection of a great power and ride its coattails (Kang distinguishes bandwagoning and hierarchy by arguing that hierarchy does not need to entail active policies to gain favor with the higher-ranked powers). The logic of bandwagoning is based on traditional conceptions of power, but it also acknowledges newer formulations of power, such as soft power. Soft power is hard to appreciate in a realist-dominated world; as Stalin famously put it, "How many divisions does the pope command?" Yet if we agree that soft power has a logic and strength of its own, then bandwagoning makes more sense; states could seek to align with another power because of ideological affinities or the convergence of values. Of course, they could balance against another state, but "the glue" in the relationship would reflect more than calculations of national interest and measurements of hard power. It should be clear at this point that the neat distinctions between balancing and bandwagoning quickly blur in the real world. For example, is Australia's strengthening of its alliance with the United States evidence of balancing against an external threat or of bandwagoning with Washington?

Recently, another model of international relations has been proposed for the Asia-Pacific: the "concert of powers." Its primary proponent is former Australian defense official Hugh White, and he has created considerable debate with his call for an order modeled on the nineteenth-century Congress of Vienna. White believes that US hegemony is Australia's best foreign policy option, but current trends make it unsustainable. Given China's history, pride, and apparent trajectory, he believes that Beijing will not accept a subordinate position in a United States–dominated and –ordered regional system. To head off the inevitable clash—based on power transition theory—he believes that the United States and other nations should move to accommodate Chinese demands by forging a new regional order that sets the United States, China, India, and Japan as the leading powers.[21]

White's theory is controversial, not least because he believes that Australia should be bending more toward China. He claims that "a country that wants to benefit from China's unique economic opportunities must . . . take careful account of China's political and strategic interests."[22] (This sounds a lot like bandwagoning.) Responses to his proposal have been fierce. Critics challenge his assumption that China's growth will continue unchecked, or counter that balancing is to be preferred when facing a revisionist power.[23] Implicit in their opposition seems to be the fear that White's logic will create a momentum of its own, making bandwagoning more acceptable.

There are other, more compelling, critiques. First, it is unlikely that any of the governments White identifies would acknowledge being part of such a concert or even wanting the responsibility. Second, and perhaps more important, other regional countries are not likely to acquiesce to such an arrangement. The international system is much more democratic than it was in the nineteenth century, and smaller countries have options other than acquiescence to a bigger, more powerful state. The Association of Southeast Asian Nations (ASEAN), which was formed in 1967, is an attempt by smaller states to join together to check the ambitions of the great powers in their region. Its insistence on "ASEAN centrality" when it comes to any regional institution building asserts its primacy in the region. While the ASEAN's ambitions outpace its capacity, its existence is a speed bump when other powers try to throw their weight around.

Conclusion: Yes, We Can

While some assert that the United States is a bullying superpower, this characterization is a caricature. The hallmark of US foreign policy in the post–World War II era has been the creation of an international order that limits the prerogatives of all nations, itself included. The United States has pressed for articulation of international rules, norms, and principles, and for the establishment of institutions that will check the exercise of raw power. As this order has evolved, it has encouraged democracy to flourish and lifted hundreds of millions of people out of poverty, affording them dignity and the opportunity to assert themselves and prosper.

Of course, the United States has benefited greatly from this order; it is in the country's national interest. But so too have billions of others. It is this history and this leadership that substantiate the claim that the United States can and will share power as China rises—as long as this emergence does not threaten other regional states and the regional order that has allowed them all to prosper and grow.

Key Points

- The United States is clearly a great power. Despite a small relative decline, America's economic, military, and soft power capacity remains substantially larger than that of any other country in the world.
- It has become clear that great power status "ain't what it used to be"; not only is it increasingly difficult for great powers to dictate outcomes in international affairs, but middle and smaller powers also may have increasing influence in Asia-Pacific affairs.
- The US "rebalance" to the Asia-Pacific is an acknowledgment of the shift in the global center of gravity toward the Pacific. It also involves the reordering of priorities within the US foreign policy tool kit. Diplomacy and economic cooperation are primary, with military aspects less significant under the conception of the Obama administration. The essence of the rebalance will continue, even if this particular term is dropped when President Obama leaves office.
- The US-China relationship is critical for a peaceful, stable, and prosperous Asia-Pacific region, yet though the two countries' national interests at times overlap, their priorities frequently do not.
- Washington and Beijing have struggled to find a new framework for their relationship that satisfies not only their interests but also those of the wider Asia-Pacific.
- Scholars debate the extent of balancing and bandwagoning occurring in the region; however, these concepts help explain some of the security changes occurring.
- While conflict remains remote, the risk is there. Approaches such as hierarchical systems or concerts of power have been raised as a solution by some scholars but seem unlikely.

Questions

1. What makes a state a "great power"? Is the United States a great power?

2. Why is the United States considered important to the post–World War II development of the Asia-Pacific?

3. What does the Obama administration mean by its policy of a "pivot" or "rebalance" to the Asia-Pacific?

4. What do the terms "balancing" and "bandwagoning" mean? Is there evidence that these concepts are guiding state behavior in the Asia-Pacific today?

5. Name some of the main points of tension in the US-China relationship.

Guide to Further Reading

Clinton, Hillary. "America's Pacific Century: The Future of Politics Will Be Decided in Asia, Not Afghanistan or Iraq, and the United States Will Be Right at the Center of the Action." *Foreign Policy* 11 (October 2011), http://www.foreignpolicy.com/articles/2011/10/11/americas_pacific_century.
   This essay by then–US secretary of state Hillary Clinton lays out the Obama administration's approach to the Asia-Pacific. Along with President Obama's speech to the Australian Parliament a month later, it is seen as a formal expression of the administration's policy of pivoting or "rebalancing" to Asia.

Kaufman, Stuart J., Richard Little, and William C. Wohlforth, eds. *The Balance of Power in World History*. New York: Palgrave Macmillan, 2007. http://dx.doi.org/10.1057/9780230591684.
   The concept of a balance of power is well known to students of European history. This book explores how it has operated (or not occurred) in other periods in time and places around the world.

Minxin, Pei. "How China and America See Each Other." *Foreign Affairs*, March–April 2014. https://www.foreignaffairs.com/reviews/review-essay/how-china-and-america-see-each-other.
   Perhaps no relationship is more important for the future of security in the Asia-Pacific than that between the United States and China. Minxin Pei is one of the United States' foremost China watchers.

Obama, Barack. "Remarks by President Obama to the Australian Parliament," November 17, 2011. http://www.whitehouse.gov/the-press-office/2011/11/17/remarks-president-obama-australian-parliament.
   This speech by President Obama in 2011 is seen as one of the clearest articulations of the justifications for and logic of the US rebalance to the Asia-Pacific.

White House. *National Security Strategy*. Washington, DC: White House, 2010. http://www.whitehouse.gov/sites/default/files/rss_viewer/national_security_strategy.pdf.
   All US administrations lay out their security policies in National Security Strategies. The Obama administration's first strategy notably focused on domestic challenges, especially economic ones. This was the administration's focus as it developed the rebalance to Asia.

White, Hugh. *The China Choice: Why We Should Share Power*. Oxford: Oxford University Press, 2013.
   This book by an Australian academic and former defense official has been hotly debated for its argument in favor of a concert-of-power approach to the US-China relationship.

Notes

1. Elaine Sciolino, "Madeleine Albright's Audition," *New York Times*, September 22, 1996.
2. Terry Atlas, "Kerry Fights Myth of US 'Fading' under Obama," Bloomberg News,

January 24, 2014, http://www.bloomberg.com/news/2014-01-24/kerry-fights-myth-of-u-s-fading-globally-under-obama.html.

3. Alan J. P. Taylor, *The Struggle for Mastery in Europe 1848-1918* (Oxford: Clarendon Press, 1954), xxiv.

4. Joseph S. Nye Jr., "Soft Power," *Foreign Policy* 80 (Autumn 1990): 153–71.

5. Aaron Friedberg, *A Contest for Supremacy: China, America, and the Struggle for Mastery in Asia* (New York: W. W. Norton, 2011).

6. Yes, the United States is also charged with violating many of these principles, but, critically, it continues to be measured against them. The United States' stature and status will be genuinely threatened when its critics give up on it and quit using these benchmarks.

7. White House, *National Security Strategy* (Washington DC: White House, 2010), http://www.whitehouse.gov/sites/default/files/rss_viewer/national_security_strategy.pdf. The Obama administration repeated core elements of this document in the second *National Security Strategy*, which it published in February 2015; see White House, *National Security Strategy* (Washington, DC: White House, 2015), http://www.whitehouse.gov/sites/default/files/docs/2015_national_security_strategy.pdf.

8. Many elements of the rebalance were in place before the Obama administration. The growing emphasis on Asia goes back at least two decades. There are more continuities than breaks with previous policy, a source of no small amount of teeth-gnashing among Republican makers of foreign policy. Product differentiation is more at work here than genuine differences in strategy.

9. Thomas Donilon, "President Obama's Asia Policy and Upcoming Trip to the Region," speech, Washington, November 15, 2012, http://csis.org/files/attachments/121511_Donilon_Statesmens_Forum_TS.pdf.

10. Hillary Clinton, "America's Pacific Century," *Foreign Policy*, October 11, 2011, http://www.foreignpolicy.com/articles/2011/10/11/americas_pacific_century.

11. This ordering of priorities is especially clear in the Obama administration's second *National Security Strategy*; see White House, *National Security Strategy* (2015).

12. Wayne M. Morrison, *China-US Trade Issues* (Washington DC: Congressional Research Service, 2014), 15.

13. Ministry of Foreign Affairs of the People's Republic of China, "Yang Jiechi's Remarks on the Results of the Presidential Meeting between Xi Jinping and Obama at the Annenberg Estate," press release, June 9, 2013, available at http://www.fmprc.gov.cn/mfa_eng/.

14. White House, *National Security Strategy* (2015), 24.

15. See, e.g., Adam P. Liff and G. John Ikenberry, "Racing toward Tragedy: China's Rise, Military Competition in the Asia Pacific, and the Security Dilemma," *International Security* 39, no. 2 (Fall 2014): 52–91; and Thomas J. Christensen, "China, the US–Japan Alliance, and the Security Dilemma in East Asia," *International Security* 23, no. 4 (Spring 1999): 49–80. A view that challenges claims that China is a newly assertive power is Alastair Iain Johnston, "How New and Assertive Is China's New Assertiveness?" *International Security* 37, no. 4 (Spring 2013): 7–48.

16. Minxin Pei, "How China and America See Each Other: And Why They Are On a Collision Course," *Foreign Affairs*, March–April 2014, https://www.foreignaffairs.com/reviews/review-essay/how-china-and-america-see-each-other.

17. Donilon, "President Obama's Asia Policy."

18. Robert Zoellick, "Whither China: From Membership to Responsibility," speech, National Committee on US–China Relations, New York, September 21, 2005, http://2001 -2009.state.gov/s/d/former/zoellick/rem/53682.htm,

19. Chen Xiangyang, "The 'New Type of Major Country Relationship' Is in Need of Support," *China–US Focus*, February 25, 2014, http://www.chinausfocus.com/foreign-policy /the-new-type-of-major-country-relationship-is-in-need-of-support/.

20. David C. Kang, "Stability and Hierarchy in East Asian International Relations, 1300– 1900 CE," in *The Balance of Power in World History*, ed. Stuart J. Kaufman, Richard Little, and William C. Wohlforth (New York: Palgrave Macmillan, 2007), 199.

21. Hugh White, *The China Choice: Why We Should Share Power* (Collingwood, Australia: Black Inc., 2013).

22. Hugh White, "The Limits to Optimism: Australia and the Rise of China," *Australian Journal of International Affairs* 59, no. 4 (2005): 469–80, at 470.

23. See "Commentary on Hugh White's China Choice," special edition of *Security Challenges* 9, no. 1 (2013), http://www.regionalsecurity.org.au/Resources/Files/SC9-1.pdf.

2 Is China an Asia-Pacific Great Power?

Lowell Dittmer

Reader's Guide

This chapter explores the rise of the People's Republic of China (hereafter, China) and asks whether China meets the criteria for great power status and, if so, what type of great power, as set out in the book's introduction. It examines how China's role in the Asia-Pacific is changing and China's views about the regional order, particularly its relations with the United States. It also considers the potential points of tension, and opportunities for cooperation, between the two powers. It then looks at China's changing role in the region. The chapter highlights the nature of China's approach to regional intergovernmental organizations and to key regional actors. Finally, it turns to China's role vis-à-vis Asia's current leading flashpoints, where rising tensions are most apt to trigger broader conflict.

Introduction

This chapter is focused on the rise of the People's Republic of China (hereafter, China). China has, since the victory of the communist revolution in 1949 (referred to in China as "Liberation"), and more specifically since the advent of the "reform and opening" policy in late 1978, made an economic ascent that is exceptional even by East Asian standards (table 2.1). And yet China has had an oddly quizzical attitude about whether it has become a "great power." The purpose of this chapter is to answer five key questions. First, is China a great power? Second, how has China's relationship with the United States—the world's "sole superpower"—changed? Third, how has China's role in the Asia-Pacific shifted in the course of its rise? Fourth, what is China's stance toward the current regional security architecture? And fifth, what are the main points of tension, or "flashpoints," in Asia and how does China relate to them?

Table 2.1 The People's Republic of China: Key Facts

Form of government	Communist
Gross domestic product (GDP)	$10,380.4 billion (2014)
GDP (purchasing power parity)	$17,617.3 billion (2014)
Population	1367.8 billion (2014)
Military spending[a]	$216.4 billion (2014)
Military personnel[b]	2,993,000 (2013)
Aircraft carriers	1
Treaty ally in Asia	North Korea

[a]Stockholm International Peace Research Institute (SIPRI), *SIPRI Military Expenditure Database* (Stockholm: SIPRI, 2015), http://sipri.org/research/armaments/milex/milex_database.
[b]World Bank, *Armed Forces Personnel, Total* (Washington, DC: World Bank, 2015), http://data .worldbank.org/indicator/MS.MIL.TOTL.P1.
Source: Unless otherwise specified above, data are from the Australian Department of Foreign Affairs and Trade, *China Country Fact Sheet* (Canberra: Commonwealth of Australia, 2014), http://dfat .gov.au/trade/resources/Documents/china.pdf.

The Quandary of National Identity

With the world's largest population (1.3 billion) and third-largest land area, China has long boasted all the qualifications to become one of the world's great powers. Moreover, although China remained geographically isolated from the mainstream of Western "world" history until the modern era, it has long deemed itself a major power. From 1500 to 1800, China accounted for an estimated 22 to 33 percent of global **gross domestic product** (GDP).[1] Only since the second half of the nineteenth century did China's share of global GDP begin to decline, reaching a nadir of 4.5 percent in 1950. Chinese public opinion has been acutely aware of this decline, defining it as a "century of humiliation" and blaming it squarely on Western and Japanese imperialism. The Chinese communist revolution was conceived as the antidote to this humiliation, seen when Chairman Mao Zedong triumphantly declared at Tiananmen Square in 1949 that "China has stood up."[2] Under these circumstances, one would certainly expect this revolutionary new nation to aim to become a great power. And yet Chinese leaders have repeatedly vowed over the years that China would never become a hegemon, or dominant power—not even after it had fully developed. Mao's successor, Deng Xiaoping, encouraged his listeners to criticize China if it ever did so.[3]

Why? There are at least three reasons. First, the Chinese word for hegemon, *badao,* has at least since Confucius (551–479 BC) had morally negative connotations lost in the English translation. True, the first generation of revolutionaries expressed contempt for Confucius (Mao said he hated him since the age of eight), but this aspect of Confucian thinking seems to have survived. Second, the **Chi-**

nese Communist Party (CCP) was a revolutionary party, viscerally opposed to all entrenched authority, identifying strongly with the oppressed and exploited classes and focused on international revolution. Any ambition to achieve great power status was ideologically suspect; the correct position was that the revolution could succeed only when it succeeded for all—particularly the "international proletariat" of new developing countries. That this was sincerely believed by these lifelong revolutionaries is amply testified by their deeds: The national goal of reuniting with Taiwan was sacrificed in 1950 in favor of an intervention to rescue Kim Il-sung from otherwise certain defeat in the Korean War. Budget resources were devoted to revolutionary insurgencies abroad, even as China's own population was starving to death in the wake of the catastrophic **Great Leap Forward** (1958–60). Finally, Deng Xiaoping, the father of reform and opening, in the early 1990s gave a famous twenty-four-character admonition to his colleagues on how to cope with Western sanctions after the crackdown at Tiananmen: "Hide your brilliance, bide your time" (*taoguangyanghui*).

Nationalism thus had to emerge in China in the form of claims to represent the only ideologically correct, pure form of internationalism. It first took the faintly patricidal form of a split with the Union of Soviet Socialist Republics (USSR), the first successful revolutionary power and erstwhile patron of the nascent CCP. Very early in this schism, Mao announced: "Comrade Khrushchev has told us the USSR fifteen years later will surpass the United States of America. I can also say, fifteen years later, we may catch up with or exceed the UK."[4] On December 2, 1957, Liu Shaoqi reiterated this aim at the Eighth National Congress of Chinese Trade Unions, and the slogan of "surpassing Britain and catching up with the United States" gained wide currency during the Great Leap Forward. Mao's frequently repeated words have since been internalized as the ultimate goal of China's developmental trajectory. During Mao's lifetime this goal was at times second to the priority the chairman placed on keeping the revolutionary spirit alive through aid to "National Liberation Wars."[5] But with the advent of "reform and opening" at the third plenum of the Eleventh Party Congress in December 1978, Mao's revolutionary agenda was quietly discarded. While Mao had made a historic contribution, the new leadership decided that his attempt to continue the revolution indefinitely was incompatible with his other ambition—to make China great. World revolution gave way to a highly focused policy subordinating everything to a rapid increase in the "forces of production," as measured mainly in GDP growth. All political objectives were subsumed by "stability above all" (*wending yadaoyiqie*).

This policy has proven highly effective, with the result that by the second decade of the twenty-first century China has, by all objective indicators, clearly become a great power economically, militarily, and politically. In the last twenty years, China has risen to become the second-largest economic power in the world. Its GDP surpassed Canada's in 1993, Italy's in 2000, France's in 2005, the

United Kingdom's in 2006, Germany's in 2008, and Japan's in 2010. In 2012 it surpassed the United States as the world's biggest trading nation. Since the advent of reform, China's economy has grown more than a hundredfold, at an average annual rate of nearly 10 percent. By October 2014 China's foreign exchange reserves reached $3.89 trillion, the largest cache ever accumulated, and its 2014 GDP (as measured in purchasing power parity) surpassed that of the United States.

As China's economy has grown, its military prowess has increased as well, initially lagging behind economic development (during the Maoist era the People's Liberation Army, or PLA, became technologically obsolescent, and military modernization was not prioritized until 1989), then exceeding it. From the mid-1990s to the late 2000s China's share of global wealth doubled while its military budget grew around sixfold. According to official Chinese figures, annual defense spending in China rose from $14.6 billion in 2000 to almost $132 billion in 2014, second only to that of the United States.[6] Washington still spends many times as much on defense, but if trends continue, China could achieve military parity with the United States in fifteen to twenty years.[7] The pace of China's development of technologically advanced weapons has continued to surprise intelligence analysts, with China developing stealth fighter jets and antiship ballistic missiles earlier than expected, and it is now experimenting with hypersonic glide missiles. Such weapons are designed to deter US forces offshore (as in the case of war in the Taiwan Strait). But a great power must also have "force projection" capability. This China clearly has in the region, with the largest (though not necessarily the strongest) navy and air force in Asia and a minimal nuclear deterrent of survivable intercontinental ballistic missiles. Though the PLA has heretofore made few efforts to "project" its force internationally, its **doctrine** and mission are presently limited to defending its peripheral borders and claims.

Chapter 1 defined a great power as a state that is able—and is recognized as able—to exert influence around the world, involving the possession of global interests and the means of protecting those interests. In this respect, China's national identity has evolved. Whereas during the Maoist period China's own estimate of its international impact outpaced its economic and strategic capabilities (as did its estimate of the potential for the world revolution it aspired to lead), in the reform era self-estimates have tended to lag capabilities. During the 1978–2008 period China was sometimes accused of being a "free rider," joining many international institutions and taking full advantage of what they had to offer while making few contributions. Not until the early 1990s did the Chinese leadership concede that China was a "responsible great power," in the context of a call for "peaceful rise" or "peaceful development." Since 2008 China has publicly acknowledged its arrival as a great power "of a new type." This became official during President Xi Jinping's summit with US president Barack Obama at the Sunnylands Retreat in Southern California in early June 2013. Here Xi introduced "China's new model for great power relations," meaning no war and mutual respect for "core interests."[8] China has hence become more willing to use its UN Security Council veto in recent years, and it has successfully pressed to increase its voting rights in the World Bank and International Monetary Fund, where it also suggested replacement of the dollar with the yuan as the international reserve currency.[9] China's state-owned enterprises have aggressively pursued natural resource assets abroad, particularly in Africa and Latin America, supported by Chinese loans, developers, and diplomatic charm offensives.

In addition to its supreme military, economic, and political power, a great power must have the vision to exert international leadership. This China certainly had during the Maoist era; the problem was a dearth of loyal revolutionary followers overseas. After a period of domestic focus and international passivity, China under Xi Jinping has again begun to project such a vision—one quite compatible with Chinese foreign policy interests but designed to have wider appeal. This vision involves a fierce critique of American leadership as intrusive, hypocritical, chaotic, and ineffectual. It is one of a multipolar world without superpowers, without security alliances and blocs, in which nations live in "harmony" based on mutually compatible interests ("win–win") and respect for sovereignty under the loose supervision of the UN. Each "pole" (China is tacitly envisaged as the leader of the Asian pole) would comprise a "community of shared destiny" (*mingyungongtong*), in which economically integrated members would support and defend one another from outside threats and "intruders," while managing internal threats together through collaborative and cooperative mechanisms: "Asians have the capacity to manage security in Asia by themselves"[10] The developing world will play a much more central role in the China dream of a new world.

China and the Superpower

The relationship between China and the United States is that between an incumbent and a rising power. Some theorists of international relations view this as a **power transition**—when a rising power overtakes and surpasses an established hegemon—and argue that it is a situation particularly susceptible to the outbreak of war. Yet neither country wants war—particularly not China, whose military capability still lags that of the United States. Fortunately, the probability of war accompanying power transition is only theoretical. Although war accompanied the rise of Germany and Japan in World War II, the United States overtook and surpassed the United Kingdom without a hegemonic war. Despite the long Cold War, there was actually no direct armed conflict between the United States and the USSR, even though the latter was viewed as a "superpower" and peer competitor.

Hegemonic war is at this point only a theoretical propensity, whereas the empirical reality is multifaceted. On one hand, the two economies have become closely linked by trade, investment, tourism, and educational exchanges, all of which would be jeopardized by hostilities. Toward the end of the Cold War, the two were strategically synchronized by joint opposition to the USSR, and strategic cooperation has to some degree continued since. Most notably, Beijing convened the Six-Party Talks in the 2000s to negotiate (unsuccessfully) a solution to the North Korean nuclear crisis. There are few specific conflicts of interest between them. Notwithstanding the hostile rhetoric of their media, China's fascination with the United States is belied by the direction of outgoing foreign direct investment, immigration, and capital flight, while China's almost romantic appeal to Americans has extended across the political spectrum, from Richard Nixon to Bill Clinton.

On the other hand, Beijing and Washington represent different ideological identities, both with messianic claims to world order. Since the 1989 Tiananmen crackdown, the United States has been a leading critic of Chinese human rights policies. Bilateral trade is not only large but is also perpetually imbalanced in China's favor. In 2012 alone, China amassed a surplus of $231 billion with the United States, accounting for 32.5 percent of the US trade deficit. This surplus is usually attributed to the migration of American manufacturing industry from the United States to China's cheap and politically disciplined labor pool, resulting in a flood of imports from foreign-invested enterprises at the expense of American jobs. Finally, China's announcement of its "core interests" in the late 2000s is problematic for the United States, not only due to its lingering attachment to Taiwan but also because the United States has alliance commitments to Japan and the Philippines, both of which have maritime territorial disputes with China. The United States also has asserted a commercial (and military) interest in continuing freedom of navigation in the seas surrounding China. To some extent, the two states are talking past each other, for though realist China defines its core interests in terms of sovereignty and spheres of influence, the liberal United States defines its interests in terms of rules and procedures (e.g., see the Taiwan Relations Act).

The official US attitude toward China's rise has been to welcome it, insisting that US trade and investment contributed substantially to that rise. But that welcome is tainted in Chinese eyes by an insistence that China conform to rules of the international arena made without China's participation or consent. Americans worry about China's ambition to revise the international order to fit Chinese interests, not only because of "power transition" anxiety but also because of repeated Chinese attempts to exclude the United States from their new world whenever diplomatically feasible. Relations have tended to oscillate from warm to cool, depending on the US electoral cycle (the American electorate has an anticommunist bias) and Chinese behavior. While US presidents Reagan, Clinton, and George W. Bush had relatively cool views of China at the outset of their terms that later defrosted, President Obama began with overly optimistic expectations, only to see them dashed by the caustic Chinese reaction to his perceived failure to respect China's core interests. This perceived core interest incongruence lies at the crux of the strategic mistrust that animated the "pivot" or "rebalance" rolled out in 2011–12 by President Obama and Secretary of State Hillary Clinton.

Although this foreign policy initiative includes an economic component in the form of a multilateral free trade association (the **Trans-Pacific Partnership**, from which China is excluded), the main thrust has been strategic. The pivot involves asserting that the United States has a "national interest" in the peaceful settlement of maritime disputes. It also shifts more US naval and air forces to the Pacific (including a new concept designed to penetrate China's **anti-area / access denial** (commonly abbreviated A2/AD) strategy, and deploying an outpost of Marines to Darwin, Australia).[11] The Chinese media response has been scathing (the diplomatic response has been a bit milder), spurning American denials that the pivot is directed against China. Some scholars see the pivot as having directly harmed US-China relations and as having led to more confrontational behavior by China.[12] Those nations bruised by Chinese assertiveness—such as Japan, Vietnam, and the Philippines—have been more supportive but skeptical about whether the initiative will be sustained amid American political deadlocks, a declining growth rate, and double deficits.

It would appear from a number of indicators that China believes great power, or "polar," status implies domination of its Asian periphery, which is incompatible with the current US position in East Asia. Beijing is aware that its power is not yet equal to that of Washington and that it would be folly to provoke war with a superior adversary. If it is to achieve its goal without precipitating a hegemonic war that would be destructive to both countries, it would be prudent to remain friends with the United States—at least until the power transition is complete. This would defer any strategic challenge until China becomes strong enough to win—preferably without fighting. China's pursuit of this path is implicit in the concept of "peaceful rise." Recently China has, however, grown more confident of its military superiority over many of its neighbors and of their asymmetric economic dependency on it. Beijing has become more willing to use this asymmetry to assert its dominant role in the region. Thus far, Beijing's more assertive policies have been aimed merely at evicting perceived trespassers from its claimed sovereign territory (even if that territory is within said trespassers' Exclusive Economic Zones). These practices are explored in more detail in chapter 7. Meanwhile, China hopes to sustain cordial relations with the United States, which in Chinese eyes has no territorial stake in the game and hence will not risk getting too deeply implicated. Thus relations remain reasonably cordial, while at the same time an underlying **national identity gap** poses a tacit challenge to the US position that is not likely to clearly emerge until the power asymmetry is fully reversed. That said, the Chinese military has a cherished belief, stemming from its revolutionary heritage, that it can prevail against superior adversaries, that "if you don't hit them they won't fall," and that it is sometimes necessary to strike first with courage and boldness to teach enemies a "lesson."

China's Changing Role in the Region

During the Maoist period China viewed itself as a revolutionary vanguard for the new nations then emerging from colonial domination. It promoted wars of national liberation in Vietnam, Burma, Laos, Malaysia, North Korea, India, Indonesia, and the Philippines. In the Vietnam War, between 1965 and 1969, the peak years of military action, Chinese PLA's Anti-Aircraft Artillery batteries allegedly shot down 1,707 US warplanes and damaged another 1,608.[13] With the rise of Deng Xiaoping, China's support for revolutionary movements in Asia came to an end. In accord with its new "peace and development" line, China shifted to the support of "good-neighbor **diplomacy**" (*mulinwaijiao*), with all Asian countries seen to be useful to Chinese economic modernization, also phasing out its former principled opposition to both "hegemonic" superpowers, the USSR and the United States. There were two major exceptions during the late twentieth century: Vietnam, against whom China fought a short, brutal war in February–March 1979 with equivocal results (China suffered prohibitive casualties, and Vietnam did not immediately withdraw from Cambodia); and India, with which China has not reached a border settlement—alone among China's Asian land neighbors. The Tiananmen massacre in 1989 escalated China's emphasis on "neighborly diplomacy," as most developed Western powers imposed diplomatic and trade sanctions and China invited its Asian neighbors into the economic vacuum.

While China's rhetorical emphasis on peace and development has continued to date, its rise to self-proclaimed great power status at the end of the 2000s has revived some regional concern. There are at least two reasons for this. First, reflecting the speed and magnitude of its economic rise, China's military buildup has outpaced the budgetary capabilities of all its neighbors, including Japan. China now has the second-largest military budget in the world, and the largest in Asia. Second, although China's maritime territorial claims have been on the public record for some time, since its articulation of the concept of "core interests" in the late 2000s Beijing has begun enforcing these claims more forcibly. This has led to a number of confrontations between Chinese coast guard and fishery police vessels and smaller Asian claimants, against which Japan, Vietnam, and the Philippines have protested most vociferously. In the case of Japan, this has resulted since 2012 in a popular boycott of Japanese imports in China, which has affected the terms of trade and resulted in growing popular antagonism on both sides. In Southeast Asia, it has resulted in a two-level game in which still-rising trade, investment, and flowery rhetoric have been accompanied by Chinese regulatory rulings and warnings (haphazardly enforced) reaffirming exclusive Chinese sovereignty claims. Negotiations on a code of conduct in the South China

Sea resumed in early 2013, with no early resolution in sight. Meanwhile, headlines to the contrary, arms spending in Asia has not kept pace (either absolutely or relative to GDP) with Chinese arms spending, which rose eightfold from 1992 to 2012.[14]

Regional Architecture

Since 1949 China has had two experiences with multilateral governmental organizations. The first, from 1949 until the late 1970s, was with the International Communist Movement (ICM), an organization headquartered in Moscow and dedicated to the global victory of the communist revolution. China was initially an enthusiastic participant in the ICM and was generally deemed its most important member aside from the USSR, representing the first exemplar and vanguard of a revolutionary strategy focused on the newly emerging nations of the Third World. In its role as communist emissary, China gave articulate voice to the aspirations of the developing world and supported Third World or nonaligned **intergovernmental organizations (IGOs)**. China's participation in the ICM turned out to have in effect been a disastrous mistake, damaging both to the movement (which it split) and to China itself. Its bitter public exit from the ICM drove China into an ideological and developmental dead end, culminating in the Great Leap Forward and the **Cultural Revolution**. It was during this period that China also adopted an extreme but unrealistic foreign policy of opposing both superpowers. Its rhetorical identification with the developing world was, however, somewhat more successful and would continue, building an international base for its 1971 entrée into the United Nations.

After its exit from the ICM, China was somewhat slow to join noncommunist IGOs for three reasons: first, the bitter aftertaste of China's recent experience with communist IGOs; second, there was a lingering communist suspicion of IGOs as puppets of bourgeois superpowers; and third China had made a point of adopting an "independent foreign policy" and joining no blocs or alliances. This was linked to China's innovative "new security concept," first introduced in the late 1990s. This approach argues that the Cold War mentality of competing and antagonistic blocks is outdated and that more informal and noncommittal arrangements can facilitate adequate security. China nevertheless secured admission to the UN in 1971, displacing the Republic of China / Taiwan, and soon after joined the alphabet soup of UN-affiliated agencies (e.g., the Food and Agriculture Organization, the United Nations Conference on Trade and Development, the International Monetary Fund, the World Bank, the World Trade Organization, UNESCO, and the International Atomic Energy Agency). Most significant of these was the World Trade Organization, which removed the threat of "most favored nation" trade discrimination and greatly facilitated China's export-oriented growth strategy.

Meanwhile, the construction of a regional multilateral architecture was only just beginning in Asia, split as it had been along various ethnoreligious and political lines. The two principal architects have been the Southeast Asian developing nations and former colonies that organized the Association of Southeast Asian Nations (ASEAN) in 1967 to protect their interests through consensual unity; and the United States, the dominant power in the region since World War II. ASEAN was founded in 1967, expanding after the Cold War to include all ten Southeast Asian nations, as is discussed in chapter 11. It became the hub for the creation of multilateral free trade agreements (with China, Japan, and South Korea, and within ASEAN), and various noncommittal but inclusive security forums, such as the **ASEAN Regional Forum**, the **East Asia Summit**, and foreign and defense ministerial meetings. The **Asia-Pacific Economic Cooperation** forum (APEC) was formed in 1989 at Australia's initiative, linking Asian and eastern Pacific countries (e.g., Canada, Mexico, and Chile) in a focus on mutual tariff reductions. The Asian Development Bank was launched under Japanese patronage in 1966. China joined APEC in 1991, joined the ASEAN Regional Forum in 1994, initiated the ASEAN+3 talks (including Japan and South Korea) in 1997, and in 2010 formed the world's largest preferential trade agreement, the ASEAN-China FTA. This soon enabled China to displace the United States as ASEAN's largest trade partner. China has also become the second-largest source of foreign direct investment in the region after the United States. In 2001 China founded the Shanghai Cooperation Organization, including Russia, Kazakhstan, Kyrgyzstan, Tajikistan, and Uzbekistan; India, Pakistan, Iran, and Mongolia later joined as observers. In 2003 China initiated the Six-Party Talks to cope with incipient North Korean nuclear proliferation, an effort that managed the crisis without, however, resolving the underlying problem. The year 2014 saw something of a breakout in Chinese international organizational initiatives, as Beijing founded two banks, the BRICS New Development Bank (the BRICS are Brazil, Russia, India, China, and South Africa—i.e., the world's largest fast-growing countries) and the Asian Infrastructure Investment Bank, providing the seed capital for both. Beijing also initiated the 21st Century Maritime Silk Road, with China-built regional container ports (Hambantota and Colombo in Sri Lanka and Gwadar in Pakistan). It has also developed high-speed railways and highways crisscrossing the Asian mainland in a "Silk Road Economic Belt," involving energy pipelines (through Burma/Myanmar, Central Asia, and now Russia) and other infrastructure. Beijing's lavish funding of these initiatives (loans, not grants, which Beijing can well afford with its huge foreign exchange reserves) is shrewdly designed to pull the periphery into the Chinese economic locomotive while at the same time stimulating demand for the Chinese export sector, where growth has been flagging.

Since becoming more actively engaged in multilateral institution building, China has shifted from passive adaptation ("free rider") to molding these institutions to fit its own priorities. China's prioritization of sovereignty leads it to prefer **anarchic IGO regimes**, like ASEAN or the Shanghai Cooperation Organization. China sometimes attempts to promote its own preferred multilateral organizations and to exclude the United States and its allies, as in ASEAN+3; the Shanghai Cooperation Organization, where the US request for observer status was declined; China's proposal for the East Asia Summit; or the Regional Comprehensive Economic Partnership, which includes China but not the United States. For its part, the United States attempts to promote organizations including its supporters in the Pacific Rim, such as APEC, or the Trans-Pacific Partnership (which excludes China). At the same time, China faces a regional dilemma, as some members of ASEAN have attempted to mobilize pressure within the organization against Chinese territorial claims and in favor of a Code of Conduct, which Beijing fears may inhibit its enforcement of those claims. Beijing's growing participation in multilateral organizations thus involves both opportunities and risks; **regionalism** can be both an instrument for China to assume a more active and influential role in the region and a constraint on China's international behavior. One fascinating area of contemporary research is the attempt to understand how China's approach to regional architecture is informed by both its strategic culture and its historical experience and memory of the Chinese tribute system.[15]

Asian Flashpoints

During the Cold War, China was either the direct sponsor (as in North Korea) or a backstage supporter (as in Vietnam, Burma, Malaysia, and Northeastern India) of many of the Asian flashpoints with the potential to escalate to large-scale or nuclear war. In the 1950–53 Korean War, China sent some 500,000 "people's volunteers" to prevent the collapse of the North Korean regime after General Douglas MacArthur's surprisingly successful US amphibious counterattack. On March 2, 1969, the Chinese PLA initiated the border clash with the USSR by ambushing border guards on the Zhenbao/Damanskii Island in the Ussuri River.[16] China made vital contributions in weapons, matériel, and logistics personnel to both Indochina wars. China provided aid and support to communist insurgencies throughout Asia, and cultivated close relations with the communist parties of Japan and Indonesia as well as with the radical factions of the Indian and Nepalese parties. Conversely, China sometimes also helped to facilitate negotiations to resolve these difficulties. Examples include the 1954 Geneva talks to negotiate solutions to a divided Korea and settle the first Indochina War, and the ambassadorial talks held in Warsaw from 1955 to 1970 to discuss the repatriation of Korean War prisoners. Yet China strongly opposed peace talks between Hanoi and Washington to resolve the second Indochina War in the late 1960s.[17]

Since the end of the Cold War, China's role has become more constructive, as Beijing severed ties with any rebel groups that posed a challenge to cordial business ties with established governments. Some former problem areas—the Sino-Soviet dispute, Vietnam, Cambodia, and Hong Kong—were successfully resolved. There are, however, four situations in Asia still capable of generating a major war: North Korea, Taiwan, the East China Sea, and the South China Sea. China has an interest in all four, and it is hard to conceive of any of them being resolved without China's support.

North Korea

China's relations with North Korea have long been close. In May 1950 Kim Il-sung secretly visited Beijing to brief Mao Zedong and the Chinese leadership on his war plans, as Stalin had informed Kim that Mao's approval was sine qua non. After North Korea's invasion of South Korea, the United Nations intervened—led by the United States. Following setbacks sustained by the Korean People's Army, China entered the Korean War in support of North Korea. In addition to dispatching some 3 million Chinese People's Volunteers to Korea to fight the United Nations Command, China also received North Korean refugees and provided economic aid during the war. Following the signing of the Korean War Armistice in 1953, China and other communist bloc countries provided extensive economic assistance to Pyongyang to support the reconstruction and economic development of North Korea. Chinese troops remained to help in reconstruction until 1958. In 1961 the two countries signed the Sino–North Korean Mutual Aid and Cooperation Friendship Treaty, which was extended twice, in 1981 and 2001, with validity until 2021. It is China's (and North Korea's) only mutual defense alliance. China, however, reserves the right to make an authoritative interpretation of the "principle for intervention" in the treaty.

During the Cold War, North Korea spurned an opening to the West while balancing between China and the USSR, even after Sino-Soviet relations soured, in order to benefit economically from both. But North Korea (like China) had a **Hallstein Doctrine** (i.e., this is the only true Korea, and if you recognize the other one, we will break relations with you). When the USSR recognized South Korea in 1990, followed by China in 1992, North Korea's relations with both were temporarily upset. Pyongyang around this time seems to have accelerated work on developing its own nuclear weapons capability, despite its previous signing (at Soviet insistence) of the nuclear nonproliferation treaty. These efforts were to continue through the next twenty years, despite the United States' determined diplomatic attempts to halt them. China assisted by attempting to induce North Korea to give up its quest, particularly by sponsoring the Six-Party Talks (2003–9). But Pyongyang persevered, despite several agreements not to do so, and it conducted three nuclear tests in 2006, 2009, and 2013, and now declares itself a "nuclear weapon state." This is one of the reasons that North Korea is a flashpoint. To be sure, North Korea is not the world's only officially unrecognized nuclear weapon state—Israel also has nuclear weapons, as do India and Pakistan. But the second factor that makes North Korea dangerous is that it is unpredictable and has demonstrated a proclivity for violence, which it has repeatedly demonstrated in armed "provocations" against South Korea and the United States. For example, in 2010 a North Korea mini-submarine apparently sank a South Korean frigate, the *Cheonan*, near the maritime border, killing forty-six sailors, and North Korean artillery later shelled Yeongpyeong Island, killing four. South Korea protested and vowed to retaliate swiftly next time. The problem with retaliating is that North Korea reciprocates and then relations get worse; the problem with negotiating is that North Korea exacts a high price for agreement but then does not keep it. The Six-Party Talks foundered on this point. There has been little interest in reviving them in view of North Korea's vow that it will never relinquish its nuclear deterrent, and others' reluctance to "buy the same horse twice."

Inasmuch as China is North Korea's only ally, the ability of the other four members of the six parties (the United States, South Korea, Japan, and Russia) to ostracize North Korea economically depends upon Beijing persuading Pyongyang to comply. Worse, the other countries (including China) find it impossible to negotiate any agreement that they can count on North Korea upholding. This both enhances China's diplomatic leverage and inflates it beyond a realistic possibility of success. Beijing in turn plays a double game, serving as Pyongyang's main diplomatic conduit, largest trade partner, and aid benefactor, while assuring the antiproliferation contingent that it shares their concerns and joins their sanctions when Pyongyang defies them. Thus Beijing comes under suspicion by both sides. Americans doubt Chinese contentions of an inability to control North Korea; China provides nearly 90 percent of North Korea's energy imports, some 80 percent of its consumer goods, and 45 percent of its food. Bilateral trade between China and North Korea reached nearly $6 billion in 2011, according to official Chinese data.

Although Beijing has joined each set of sanctions, there is also some skepticism about enforcement; Chinese exports of banned luxury goods averaged about $11 million per month in 2009, and in 2013 overall trade was up more than 10 percent.[18] This prompts suspicion among those trying to induce Beijing to persuade Pyongyang: Just how committed is Beijing to nonproliferation? It would appear that there are two different sets of priorities; while the West puts nonproliferation at the top, justifying any means to that end and not really minding if the Pyongyang regime collapses, Beijing's top priority has always been stability. Nuclear weapons are a problem for the West (less so for China), but for North Korea to collapse would be disastrous for China, not only because of the refugee problem but also the possible contagion. True, Beijing has expressed increasing annoyance at Pyongyang's defiance (what other country among China's maritime neighbors dares detain Chinese fishing boats and hold them for ransom?), but their priorities are unlikely to have changed. Under these circumstances, despite

Beijing's calming influence on Pyongyang, this hot spot is likely to simmer for some time.

Taiwan

Reunification with Taiwan is very high among Beijing's priorities, an explicit one of its "core interests." First annexed from the Portuguese and indigenous Chinese outlaws in the early Qing period, the island was seized by Japan after winning the first Sino-Japanese War and remained an imperial colony from 1895 to 1945. In the settlement of the Pacific War, it was returned to the Republic of China (ROC), but the Red Army then defeated the ROC in 1949. There was no peace treaty, however, and the ROC's leadership, the Kuomintang of China (KMT) regime, insists that it has only temporarily retreated to Taiwan, reconstructing the island's economy while awaiting its chance to "retake the mainland" (*huifudalu*). Beijing, however, ignores this and claims the island as its own. After forfeiting the chance to retake Taiwan in 1950 by investing instead in the defense of North Korea, Mao moved to "liberate" Taiwan, precipitating two cross-Strait crises in the 1950s that were frustrated by American intervention. After Mao's death, Deng Xiaoping shifted from armed liberation to "peaceful reunification." In 1979 he offered "three direct links" (transportation, trade, postal), leading to peaceful reunification under "one country, two systems' with substantial **autonomy**. Taiwanese president Chiang Ching-kuo responded with "three no's."

Shortly before his death in 1988, Chiang decreed a posthumous democratic transition, which ensued quickly and quite smoothly. This has had a contradictory impact on prospects for reunification. On one hand, it gave the populace greater freedom, with the result that the business community in effect soon realized Deng's three links indirectly, through Hong Kong. Trade with the mainland began in the early 1990s, followed by investment, and grew rapidly; all attempts to contain it politically came to little avail. By 2012 China was Taiwan's leading trade partner, providing some 40 percent of its trade ($197 billion in 2012) and estimates of well over $150 billion in investment. On the other hand, democratization freed the indigenous population to express pent-up resentments and anti-mainlander sentiment (blanketing both KMT migrants and the CCP regime on the mainland) found its expression in the voting booth. With far less attachment to retaking the mainland than the KMT, this constituency was open to frank appeals for Taiwan's independence (*taidu*, or *dutai*), which became the unofficial slogan of the opposition Democratic Progressive Party. But *taidu* infuriated the mainland, which threatened to invade and passed a law (unanimously) in the National People's Congress in 2005 outlawing secession.

What has made Taiwan a flashpoint? First, the **"one-China principle"** (there is only one China, and Taiwan is a part of China), which Beijing has successfully ensconced internationally as a precondition for recognition of China, has thereby severely curtailed Taiwan's international diplomatic and commercial space. Second, China has continued to insist on its right to implement this principle by force if necessary and on its growing power to do so. The threat of force was dramatized via artillery bombardment of the offshore islands in the Mao period but discontinued with the introduction of Three Links in 1979. It was, however, revived by Jiang Zemin in 1995–96 by lobbing missiles off Taiwan's leading seaports and staging amphibious exercises along the opposite coast. Although Beijing has since been less vocally threatening (in view of the electoral backlash in Taiwan), it has continued to enhance its capability to intimidate. By the end of 2012 the PLA had more than 1,600 ballistic missiles aimed at Taiwan, and their number and quality have continued to grow. China is now far bigger and more powerful than Taiwan in both GDP and military force projection.

The United States has sought to cool this hot spot with "strategic ambiguity," refusing to say what it would do if Beijing opts to use force or if Taiwan declares independence while disapproving both, but continuing to sell defensive weapons to Taipei and declaring its interest in a peaceful solution. The third factor in this combustible mix is the continuing disinterest of the Taiwan electorate in political reunification, despite ongoing cross-Strait economic integration and growing political conciliation since the election of KMT chair Ma Ying-Jeou in 2008. The Democratic Progressive Party's electoral future is based on this majority sentiment; at the same time, any electable leadership in Taiwan must also be able to maintain cross-Strait peace with the island's main trade partner. Given that all sides are aware of the enormous stakes involved, the prospects are quite good for continued mutual incremental adjustment, albeit without any foreseeable prospect of definitive resolution; that is, *bu du, bu tong* (no independence, no unification).

The East China Sea

Tokyo first laid formal claim to the eight tiny islets Japan calls the Senkakus and China calls Diaoyu Dao (hereafter, D/S) on January 14, 1895, while still fighting the first Sino-Japanese War. Japan claimed the islands were **terra nullius** and "showed no trace of having been under the control of China."[19] Five months later, Japan and the Qing government of China signed the Treaty of Shimonoseki, ending the Sino-Japanese War, on the basis of which China ceded Taiwan to Japan, "together with all the islets appertaining or belonging to said island" (not mentioning D/S).[20] Japan had control of both Taiwan and D/S from 1895 to 1945, with no audible objection from China. In the Treaty of San Francisco in 1951, Taiwan was returned to China (in accordance with the 1943 Cairo conference), while the United States assumed temporary administrative control of the Ryukyu chain, including Okinawa and D/S. In April 1971 D/S was returned to Japanese administrative control along with Okinawa, of which it became a prefecture. Chinese official maps, school textbooks, and archival materials during this period (1895–1971) acknowledged Japan's title, referring to the islets by their Japanese name. In 1968 the UN Economic Commission for Asia and the Far East conducted a survey of the area and discovered evidence of substantial subsurface carbon deposits in the surrounding waters.

In 1971 both China and Taiwan protested the return of D/S to Japan (Beijing's claim is based on antecedent ROC claims), asserting based on historical evidence that D/S had been discovered by Chinese in the fourteenth century and controlled by them since the Qing took power, had then been annexed by Japan during the first Sino-Japanese War in an unequal treaty, and should now be returned to the rightful owner. Japan stood on international law, claiming that because no legal complaint had been raised previously, its ownership of the islets was undisputed (though, in a tacit acknowledgment of a dispute, Japan has made no use of the islets since ownership became controversial). Although both China and Taiwan protested, as China was preoccupied with domestic events at the time (the Cultural Revolution) and Taiwan had otherwise cordial relations with Japan (which recognized the ROC until September 1972), China's claim was kept alive, mostly by independent activists and scholars from Taiwan and Hong Kong, for the next forty years, with some sailing (or even swimming) to and attempting to land on the islands. Both governments tacitly rested their case until 2010, when Japanese coast guard ships briefly detained a Chinese fishing boat and crew for trespassing in territorial waters, sparking demonstrations in China until all were returned. Tensions rose again in 2012, when the Japanese Noda administration purchased three of the eight islands from their private owner in order to prevent their purchase by Tokyo mayor Ishihara Shintaro, who had been raising funds to purchase and fortify the islets ("What's wrong with Japanese protecting Japanese land?" he asked). This "nationalization" precipitated demonstrations in two hundred Chinese cities, a boycott of Japanese goods, and property vandalism in the biggest mass movement since 1989. Since January 2013 Chinese ships and aircraft have patrolled the islets over Japanese protests, a Chinese Foreign Ministry spokesperson has called D/S a "core interest," and in December 2013 Beijing declared an Air Defense Identification Zone that includes the islets, evoking immediate protests from Japan, the United States, South Korea, Taiwan, and Australia.

Despite two Sino-Japanese wars and the current territorial dispute, postwar Sino-Japanese relations were actually quite good until the 1990s. Japan was among the first developed countries to recognize Beijing (without offering war reparations but consistently giving overseas development aid to China, mostly in the form of subsidized loans) and soon developed enormous trade and investment links. Although trade has for obvious reasons declined somewhat since 2012, China is still Japan's largest trade partner. Prospects for reconciliation are unfortunately not bright in the immediate future, for China's overreaction to the Japanese government's purchase of the islets prompted an equal overreaction in Japan and the election of the nationalist prime minister Shinzō Abe. Finally, at the Beijing APEC summit in 2014, the two met with a frosty handshake and agreed to begin new bilateral talks.

The South China Sea

Although the Paracel and Spratley islets in the South China Sea were visited by Chinese and other seafarers for centuries in the premodern era, the two largely uninhabited island groups were both claimed by the French until Japan invaded the islands during World War II. When the war ended, Japan renounced its rights to the islands and title reverted to France. South Vietnam acquired sovereignty over the islands by right of cession after the French-Indochina War. Following unification of the country in 1975, the Socialist Republic of Vietnam inherited South Vietnam's title to the two island groups. Meanwhile, the ROC government in Nanjing occupied the islands that had earlier been controlled by the Japanese based on claimed historic visits and took up the same claim. China's claim is known as the nine-dotted line—originally an eleven-dotted line—named after the dashed line sketched on a map by the ROC in 1947. After the victorious Red Army took the mainland and formed China in 1949, the line was adopted and revised to nine dashes as endorsed by Zhou Enlai, the first premier of the People's Republic of China. China wrested the Paracels and part of the Spratlys from Vietnam after naval engagements in 1974 and 1988, respectively, and it seized Mischief Reef and Scarborough Shoal from the Philippines in 1995 and 2012 (map 2.1).

When the UN Convention on the Law of the Sea requested a filing of territorial claims in 2009, on May 7 China submitted the nine-dashed line map claiming some 80 percent of the South China Sea as sovereign territorial waters, precipitating immediate protests from the Philippines, Vietnam, Malaysia, and Indonesia. The chief current economic resource is fishing, though there are possibly also vast (precise estimates vary) subsurface hydrocarbon deposits. While continuing to assert the legality of its claim, China's enforcement efforts have been coercive but incremental, using only enough force to prevail, preferring to use civilian rather than naval vessels. This has the advantage of avoiding a *casus belli* but the disadvantage that rival disputants are never decisively defeated and the dispute drags on (e.g., to Chinese annoyance, Manila has taken its claim to the International Court of Justice, and it was joined in 2014 by Vietnam). Aware that the protracted dispute has been damaging to China's soft power, China's president, Xi Jinping, and its premier, Li Keqiang, launched a charm offensive with their respective appearances at the 2013 APEC and ensuing East Asia Summit meetings, visiting Vietnam and other Southeast Asian nations on the way and signing deals worth billions in joint infrastructure investments. The CCP's leadership also convened an internal policy conclave in late October 2013 focusing on strategies for further improving China's relations with neighboring states. Its territorial claims have not been rescinded, but it also agreed in 2013 to resume negotiations on a

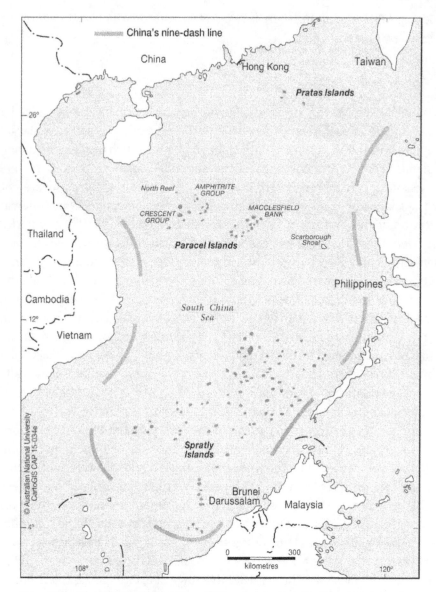

Map 2.1 The South China Sea. © The Australian National University, College of Asia and the Pacific, CartoGIS.

Code of Conduct (though progress since has not been encouraging). Scholarship on China's territorial disputes, however, gives some optimism that China is unlikely to resort to violence in territorial disputes.[21]

Conclusion

Based on the definition in chapter 1, China has definitely "arrived" as the world's newest great power. It has global interests and is developing the means to protect them. Equally important, it is aware of its arrival. Yet there is still evident uncertainty in China about how to play its new role. China deems itself to be a great power of a new type, the largest developing country and thus the natural leader of the developing world, introducing a new vision of crafting a world without superpowers, divisive alliances, and blocs based on mutual respect for sovereignty and "win–win" transactions. Beijing's visionary rhetoric is strikingly liberal, lauding "international democracy," a "harmonious world," and not only noninterference but "tolerance" (*baorong*) for all states. Yet in the pursuit of its own foreign policy interests, China has been realist, self-interested, and willing to use force or deceit when expedient. And it has been repressive domestically while being liberal abroad. In the normal course of diplomacy, this does not matter too much because other states are chiefly concerned with how they are treated here and now, and Chinese diplomats have generally been happy to negotiate mutually satisfactory deals on a quid pro quo basis. But it has given rise to a somewhat fragile foreign policy without a secure loyalty base.

Thus, in its most recent incarnation, China's foreign policy has been trying to position it more firmly as a great and benevolent power. But this has been tricky, given Beijing's vast ambitions. China aims eventually to be the unrivaled leader of the Asian region and a representative of the emerging world. To do so, it must continue to grow more rapidly than its neighbors, thereby remaining a desirable economic partner and commercial center, a natural leader, and a development role model. To continue to grow rapidly, it needs to import more raw materials and conquer more export markets. To do so, it needs to integrate Southeast Asia and Central Asia economically, dominating their trade so that their asymmetric dependency will necessitate their voluntary compliance with Beijing's political preferences. This is a "community of shared destiny." If they were to fall afoul of Beijing's preferences, they could now be easily disciplined without risk to Beijing because of the asymmetric stakes. So long as China's hypergrowth and demand pull can be maintained, it may be well worth forfeiting occasional visits by the Dalai Lama.

Aware of the tenuous economic foundation of its expanding power at a time of slowly declining GDP growth, China has recently sought to strengthen its position with a more clearly thought-out rationale for the changes in the world order it deems necessary. This schema is both a trenchant and often accurate critique of recent American foreign policy failures and a vague approximation of how a more satisfactory order might be constructed. And it has been incentivized by the promise of generous loans, investments, and mutually beneficial economic cooperation for those who join. New banks, joint projects, and international organizations have been established to ensure a lasting foundation for this new political economy. Its appeal should by no means be discounted—already there are twenty-five national subscribers to the Asian Infrastructure Investment Bank.

Key Points

- China has definitely arrived as the world's newest great power. It has substantial economic, military, political, and social power.
- China's relationship with the United States is seen as a power transition, with clear risks of conflict, especially given the different ideologies of the two countries.
- China's role in the Asia-Pacific is shifting thanks to its rise, and this is producing tension as China reasserts a position of dominance it enjoyed centuries before.
- China is increasingly willing to embrace the regional security architecture of the Asia-Pacific, though without adopting all the region's rules and norms.
- The four flashpoints of most concern where conflict could occur are North Korea, Taiwan, the East China Sea, and the South China Sea. All four involve long historical claims and memories and feature a clash between the interests claimed by Beijing and the norms and values prized by the Asia-Pacific, especially the United States.

Questions

1. If China is a great power, why has it had so much difficulty acknowledging this? Has it now clearly decided what it means for it to be a great power?

2. What are China's main objectives now that it is a great power, and what strategy does it have to achieve them? Is this a realistic strategy in your judgment? How does it affect the foreign policies of its neighbors?

3. How does China fit into the regional architecture of the Asia-Pacific? What are its major contributions to the structure of emerging Asian regionalism?

4. What opportunities and problems are created by China's rise, and what are the likely regional security implications?

Guide to Future Reading

Johnston, Alastair Iain. *Cultural Realism: Strategic Culture and Grand Strategy in Chinese History*. Princeton, NJ: Princeton University Press, 1995.
   This book traces China's "parabellum" security culture back to the Ming Dynasty, revealing a lasting historical fascination with surprise, deceit, and other recurring patterns.

Kang, David. *East Asia before the West: Five Centuries of Trade and Tribute*. New York: Columbia University Press, 2010.
   Many of China's ambitions are based upon its memory of the Chinese tribute system. This book explores how this system operated and argues that its hierarchical model helps explain the relative peace in East Asia, and how this peace could be sustained via a similar model today.

Nathan, Andrew, and Andrew Scobell. *China's Search for Security*. New York: Columbia
University Press, 2012.
> *This book provides an analysis of China's security concerns on four fronts: at home, with its immediate neighbors, in surrounding regional systems, and in the world beyond Asia. By illuminating the issues driving Chinese policy, it provides a new perspective on the country's rise and a strategy for balancing Chinese and American interests in Asia.*

Taylor Fravel, M. *Strong Borders, Secure Nation: Cooperation and Conflict in China's Territorial Disputes*. Princeton, NJ: Princeton University Press, 2008.
> *The first comprehensive study of China's territorial disputes, this book argues that China's propensity to resort to violent conflict over territory has been considerably exaggerated. China has been willing to use force to reverse a decline in its bargaining power with powerful neighbors when positioned to do so, but more typically has been willing to compromise.*

White, Hugh. *The China Choice: Why America Should Share Power*. Melbourne: Black Inc, 2012.
> *This book argues that the United States must share power with China if it is to avoid conflict in the Asia-Pacific. It is a clear, logical, and easy-to-read summary of the primary strategic choices facing Beijing and Washington.*

Notes

1. See Angus Maddison, *Contours of the World Economy, 1–2030* (Oxford: Oxford University Press, 2007).

2. David Scott, *China Stands Up: The PRC and the International System* (Milton Park, UK: Routledge 2007), 1.

3. In April 1974, in a special address to the UN General Assembly, Deng Xiaoping declared that "China is not a superpower, nor will she ever seek to be one. If one day China should change her color and turn into a superpower, if she too should play the tyrant in the world, and everywhere subject others to her bullying, aggression, and exploitation, the people of the world should identify her as social-imperialism, expose it, oppose it, and work together with the Chinese people to overthrow it." As former foreign minister Li Zhaoxing stated more recently, "China will never seek to become a hegemonic power." Jeremy Blum, quoting Li's February 12 Washington speech, *South China Morning Post*, February 13, 2014.

4. Nikita Khrushchev, *Khrushchev Remembers* (Boston: Little, Brown, 1970), 250–57.

5. Until 1964, the People's Republic of China (PRC) had been providing military hardware to some revolutionary countries, but the number jumped from a dozen to more than sixty from 1964 to 1968. By the end of 1978, more than sixty countries—including Bangladesh, Nepal, Sri Lanka, Egypt, Zimbabwe, Gabon, Sudan, Uganda, Zambia, and Somalia—had received the PRC's military aid, the most significant of which was Vietnam. Han Huaizhi and Tan Jingqiao, eds., *Dangdaizhongguojundui de junshigongzuo* (Military work of the Chinese armed forces in contemporary China) (Beijing: Chinese Social Science Academy Press, 1989), vol. 2, 578–80.

6. These are the Chinese official spending figures, which do not include such things as foreign weapons purchases and research and development. Estimates including such expenditures vary from a third to 100 percent larger. "The Dragon's New Teeth," *The Economist*, April 7, 2012, http://www.economist.com/node/21552193.

7. Peter Apps, "East-West Military Gap Rapidly Shrinking: Report," Reuters, March 8, 2011, http://www.reuters.com/article/2011/03/08/us-world-military-idUSTRE7273UB20110308.

8. China's "core interests," empirically defined, are unquestioned sovereignty over Tibet, Xinjiang, Taiwan, and, it appears, the East China Sea and the South China Sea.

9. China has vetoed four UN Security Council resolutions since 2010. All were to block international intervention in the Syrian civil war, and all were submitted jointly with (and probably at the behest of) Russia.

10. Xi Jinping, speaking at the Conference on Interaction and Confidence Building Measures in Asia, as quoted by Xinhua, May 21, 2014.

11. "Anti-access / area denial" is the Pentagon's term for a Chinese strategy designed to defend China's core interests, including Taiwan and apparently much of the South China Sea and East China Sea, and to develop asymmetric weaponry capable of deterring American intrusion into these areas.

12. Robert Ross, "The US Pivot to Asia and Implications for Australia," Centre of Gravity Series, Australian National University, March 2013, http://ips.cap.anu.edu.au/sites/default/files/COG5_Ross.pdf.

13. Han Huaizhi, *Dangdaizhongguojundui de junshigongzuo*, 552. This number has not been verified on the American side.

14. See David Pilling, "Asia Follows China into an Old-Fashioned Arms Race," *Financial Times*, April 2, 2014, http://www.ft.com/intl/cms/s/0/9d83bf62-b9b9-11e3-a3ef-00144feabdc0.html#axzz3PeXuZx00; and David C. Kang, "A Looming Arms Race in East Asia?" *The National Interest*, May 14, 2014, http://nationalinterest.org/feature/looming-arms-race-east-asia-10461?page=4.

15. For two strong introductory texts on this growing avenue of research, see Alastair Iain Johnston, *Cultural Realism: Strategic Culture and Grand Strategy in Chinese History* (Princeton, NJ: Princeton University Press, 1995); and David Kang, *East Asia before the West: Five Centuries of Trade and Tribute* (New York: Columbia University Press, 2010).

16. Yang Kuisong, "The Sino-Soviet Border Clash of 1969: From Zhenbao Island to Sino-American Rapprochement," *Cold War History* 1 (2000): 21–52.

17. Qiang Zhai, *Beijing and the Vietnam Peace Talks, 1965–68: New Evidence from Chinese Sources*, Cold War International History Project, Working Paper 18 (Washington, DC: Woodrow Wilson International Center for Scholars, 1997), http://www.wilsoncenter.org/publication/beijing-and-the-vietnam-peace-talks-1965-68-new-evidence-chinese-sources.

18. Scott Snyder, "China–North Korean Trade in 2013: Business as Usual," *Forbes*, March 27, 2014, http://www.forbes.com/sites/scottasnyder/2014/03/27/44/.

19. Ministry of Foreign Affairs of Japan, *Fact Sheet on the Senkaku Islands* (Tokyo: Ministry of Foreign Affairs of Japan, 2014), http://www.mofa.go.jp/region/asia-paci/senkaku/fact_sheet.html.

20. Uking Sun, "Backgrounder: The US Role in the Diaoyu Islands Dispute," *China Daily*, April 25, 2014, http://www.chinadaily.com.cn/world/2014-04/25/content_17465180.htm.

21. M. Taylor Fravel, *Strong Borders, Secure Nation: Cooperation and Conflict in China's Territorial Disputes* (Princeton, NJ: Princeton University Press, 2008).

3 Are India and Japan Potential Members of the Great Power Club?

H. D. P. Envall and Ian Hall

Reader's Guide

This chapter considers whether India and Japan could join the United States and China as "great powers." It argues that though both India and Japan may claim such status in the future, it is unclear whether either state actually aspires to this status and that, for the time being, both are some way from acquiring it. Both states face major domestic challenges, both have complex relationships with the United States and China, and both have political elites who are conflicted about the best direction for their foreign and security policies.

Introduction

India and Japan face some common strategic challenges—above all, they both need to manage the rise of China—but they do so with different sets of capabilities and worldviews. While many realists argue that wealthy and rising powers generally seek greater political and military power in international relations, the cases of India and Japan illustrate that liberal and constructivist arguments about the ways in which international law and institutions, on one hand, and perceptions and ideas, on the other hand, also shape the security dynamics of the Asia Pacific.

India's economy has grown since the end of the Cold War, and it has acquired nuclear weapons, but it remains a developing country. It is home to a third of the world's poorest people, it is located in a difficult immediate geographical neighborhood, and it faces many internal security challenges. Japan, by contrast, is a high-income country with an advanced economy, but one that has mostly stagnated for two decades. Its population is getting older, as fewer children are born to replace the aged, and the aged live longer. Japan has a well-equipped, professional military, but there remain strict limits on what it can spend on defense and on what it can do with its troops. And though some of India's elite aspire to the "great power club," Japan's elite is more ambivalent.

This chapter assesses the relative capabilities of India and Japan and the chang-

Table 3.1 India and Japan: Key Facts

Characteristic	India	Japan
Official languages	Hindi, English	Japanese
Population	1.26 billion (2014)	127.1 million (2014)
Land area	3.3 million square kilometers	378,000 square kilometers
GDP	$2,049.5 billion (2014)	$4,616.3 billion (2014)
Principal export destinations	United States (13.3%), United Arab Emirates, Hong Kong	United States (18.6%), China, South Korea
Principal import sources	China, Saudi Arabia, United Arab Emirates	China, United States, Australia
Number of military personnel[a]	2,749,700 (2013)	259,800 (2013)

[a]World Bank, *Armed Forces Personnel, Total* (Washington, DC: World Bank, 2015), http://data .worldbank.org/indicator/MS.MIL.TOTL.P1.

Source: Unless otherwise specified above, data are from the Australian Department of Foreign Affairs and Trade, *Japan Country Fact Sheet* (Canberra: Commonwealth of Australia, 2014), http://dfat .gov.au/trade/resources/Documents/japan.pdf; and Australian Department of Foreign Affairs and Trade, *India Country Fact Sheet* (Canberra: Commonwealth of Australia, 2014), http://dfat.gov.au /trade/resources/Documents/inia.pdf.

ing worldviews of their policymaking elites who are involved with foreign and security policy. It argues that India's rise is likely to be slower and more difficult than some enthusiasts might believe and that Japan, with its complicated history in the Asia-Pacific, remains conflicted about the kind of power it wishes to be. It argues that India's ability to shape regional order is inhibited by inherited, but still influential, attitudes in New Delhi that are skeptical about alliances, and by China's ability to dissuade New Delhi from courses of action that might not be in Beijing's interests. It argues that Japan's influence will depend on whether or how it transitions from its current role as a constrained major power, to becoming either a more modest middle power or a fully independent strategic power (table 3.1).

India

When India became independent in 1947, it had a population of about 350 million and a gross domestic product (GDP) of about $222 billion.[1] Sixty-seven years later, in 2014, India's population was over 1.2 billion and its GDP was estimated at about $2.049 trillion. But with a huge population, the bulk of which is employed in agriculture, in global terms India remains poor, with a limited capacity to maintain military forces with a regional, let alone global, reach.

India's leaders have tried coping with its weaknesses in different ways. From

1947 to 1964, under its first prime minister, Pandit Jawaharlal Nehru, India es-poused "**nonalignment**," aiming to maintain good relations with both East and West, and the developing world, but making no alliances. For Nehru, interna-tional security was best achieved by adhering to principles of "peaceful coexis-tence"—respecting the sovereignty and independence of all states, not interfer-ing in their internal affairs, and refraining from aggression. These convictions were shaken, however, by China's defeat of India in a border war in 1962 and a series of conflicts with Pakistan.

After Nehru's death in 1964, India's foreign and security policy moved toward what is known as "militant Nehruvianism." India sought what it called "strategic **autonomy**," but downplayed its earlier idealism.[2] India modernized its military, fought and won two wars against Pakistan (in 1965 and 1971), and concluded a Treaty of Friendship and Cooperation with the Union of Soviet Socialist Repub-lics (USSR), which gave it access to Soviet markets and military hardware. It also initiated a program to build a nuclear bomb, in response to China's test of a bomb in 1964. In 1974, it tested a nuclear device, but did not move to develop or deploy nuclear weapons.

The collapse of the USSR in late 1991 began the third phase of Indian for-eign and security policy. Plunged into economic crisis by the dramatic rise in oil prices caused by the 1991 Gulf War, India's government embarked on some liberalization of its heavily protected economy to try to stimulate growth, and began to build new relationships that might give it access to badly needed capi-tal. It launched a "Look East" policy, trying to improve relations with East Asian states, including Japan and the ASEAN countries.[3] But India's difficult relationship with Pakistan, which continued to sponsor insurgency in the disputed region of Kashmir, persisted, and Beijing's inexorable rise caused concern to New Delhi. India's elite realized that its economy, though improving, might take decades to catch up with China's, that its military was in desperate need of modernization, and that, unlike other regional states, like Australia or Japan, it was not protected by great power security guarantees.

To give India breathing space to solve these problems, as well as to assert its standing as a major power, its leaders ordered the testing of five further nuclear devices in 1998, declared India a fully fledged nuclear power, and moved to de-velop and deploy nuclear warheads and delivery systems (principally aircraft, eventually to be replaced by nuclear-capable missiles launched from land and sea). This move was widely condemned by other states, which were angry about the impact of India's move on the nuclear nonproliferation program, and was swiftly followed by six Pakistani tests and the announcement that Pakistan was now also a nuclear power.

In time, the tests stimulated a process of diplomatic reengagement between India and the United States, which had long been estranged from each other, but whose elites began to perceive that better relations might be in their mutual

Table 3.2 India: Key Dates

Date	Event
1947	Date of independence
1947–48	First Indo-Pakistani War
April 1965 and September 1965	Second Indo-Pakistani War
December, 3–16, 1971	Third Indo-Pakistani War
1974	First nuclear test
1998	First nuclear weapons test

interest.[4] This process intensified after the election of US president George W. Bush in 2001. Bush saw India as a "natural ally" (in the words of India's former prime minister Atal Bihari Vajpayee),[5] a democratic state with values shared by Americans, and a partner in managing China. In 2005 the United States and India concluded a deal that acknowledged India's possession of nuclear weapons and lifted restrictions on its access to civilian nuclear technology.[6]

In the mid-2000s, however, India's economic growth slowed, and some in India's elite began to express doubts about the direction of its policies, stimulating a wider debate about India's role and aspirations in the world. Some regretted the close ties forged with the United States and argued for a return to "nonalignment" or just "militant Nehruvianism." Others sought a new, more "realist" foreign policy that would entail a push for great power status and capabilities, as well as formal or informal alliances with the United States and some regional powers, including Japan, to balance China's growing influence in the Indo-Pacific region.[7] The dominant group worried about India's ability to become a great power, and recommended "hedging," not wanting to confront China directly and not wanting to tie India's fortunes to a declining United States (table 3.2).[8]

Great Power India?

India's political elite has long debated its actual and preferred status in world politics.[9] Nehru and his followers wanted India to be acknowledged as a great power, but of a new kind: a standard-bearer for more peaceful and just international relations—a "normative power" rather than a conventional one.[10] This idea of Indian "exceptionalism" persists today, even when modified by those who want India to be strong militarily as well as morally.[11] At the other end of the spectrum, there are those who want India to cast off this moralistic baggage and become a conventional, realist great power, pursuing only its national interests.[12]

These elite debates depend on differing views—and differing measures—of India's power and potential. If we take Martin Wight's realist definition of a great power—a power that has global interests and the capacity to promote and defend them, if necessary by force—then India is far from becoming one. India's economy is growing, but remains small in comparison with those of other major powers, especially in per capita terms.[13] The World Bank ranked India's GDP tenth in the world in 2012, at $1.841 trillion, but that makes its economy only about 11 percent of the size of the United States' or 22 percent of China's.[14] India's economy is roughly the size of Canada's, but supports more than thirty times the population. About a third of India's people live at or below the poverty line, and rural literacy rates, especially among women, remain poor by international standards.

India's linkages with global markets are thin; it is very dependent on imported oil and natural gas, especially from the Middle East, though its high-technology and services sectors are well-connected into Western economies. The bulk of India's GDP growth over the past twenty-five years has come from growing domestic demand for domestically produced goods and services.[15] This is a source of both strength and weakness—India was arguably able to weather the global financial crisis of 2008–9 better than China because it was less exposed to global market instability; but at the same time, its economy would likely grow faster if it had more access to foreign capital and know-how.

India has a huge number of military personnel—the third-largest globally, at 1.3 million. The army claims the biggest chunk of the defense budget, arguably starving the air force and navy of resources, and the quality of the equipment of all services is variable. Much of the hardware is aging; the bulk of India's tanks date from the 1970s, while some of its fighter aircraft were supplied in the 1960s. It has roughly 80 to 100 nuclear warheads and some systems to deliver them—at present, a mix of aircraft and short- to medium-range missiles—but its warhead designs are the least-tested of any nuclear power and it is far from realizing the delivery systems to which it aspires. India has a weak indigenous defense industry—with some pockets of real excellence, as in ballistic missiles—that struggles to produce good systems at a reasonable price. For these reasons, it has long been the biggest importer of military hardware in the world, acquiring billions of dollars of weapons from traditional suppliers, like Russia and France, and also from new ones, like Israel and the United States. But its notoriously slow—and sometimes corrupt—defense procurement system commonly delays the purchase of new equipment for years and sometimes decades.[16]

India is also beset with pressing, complex internal and external security challenges. It has faced multiple insurgencies since 1947. These include the ongoing conflict with separatists in Kashmir; clashes with various ethnic groups, like the Mizos in India's Northeast, who fought a twenty-year conflict with government forces from the mid-1960s to the mid-1980s; and more recent troubles with the Maoist guerillas known as the Naxalites, whose violence affected fully one-third of eastern India in the mid-2000s. While these conflicts have diminished in scale in the past decade, they continue to sap the resources of the state and undermine its legitimacy, especially where the actions of security forces are or have been repressive.

The instability of India's immediate geographical neighborhood, and long-running disputes with China and Pakistan over their respective borders, the treatment of minorities, and the management of resources, including water, also undermine India's ability to project power and influence beyond its borders.[17] Political crises and conflict in Burma/Myanmar, Nepal, and Sri Lanka, as well as illegal immigration from Bangladesh, compound these problems, tying up India's stretched bureaucracy and security forces and distracting them from other issues.

Finally, India lacks the political and diplomatic apparatus of even a middling state, let alone that of a great power. Foreign and security policymaking in India is concentrated in the hands of a small group—the prime minister and a few key ministers and advisers, with uniformed officers kept well away from the policy-making process for fear of military interference in civilian decision making. India continues to have one of the smallest diplomatic services of any major state—it has roughly the same number of foreign service officers as Singapore—and one of the weakest academic and think tank sectors providing advice from outside government. The few diplomats India has are high quality, but they are widely dispersed across the Ministry of External Affairs and its overseas missions, and thus are overburdened.[18]

The Rise of India and the Regional Security Order

For the most part, and in contrast to China, India's rise has been perceived with equanimity and some enthusiasm in the wider Asia-Pacific region. In South Asia and in Beijing, however, the response has been more mixed. Pakistan, above all, worries that a stronger India threatens its interests and even perhaps its integrity. Beijing protested angrily about India's nuclear tests in 1998 and refuses to acknowledge New Delhi's claim to a permanent seat on the United Nations Security Council. It has also made little effort to resolve its long-standing territorial dispute with India. There are also periodic tensions with Bangladesh and Sri Lanka, as well as fears in the small states of the region—Bhutan, the Maldives, and Nepal—that India can be overbearing and interfering. But these concerns are balanced by the hope that India's economic growth will have spillover effects and stimulate greater regional integration.

From a broad perspective, one can see that India's rise has affected the region's security order in three ways. First, its 1998 nuclear test decisively shifted the strategic balance in Asia, even if commentators argue about India's capacity to properly deter China.[19] Introducing nuclear weapons into South Asia dramatically raised the stakes of conflict with Pakistan and China, as well as providing an opportunity for a partial rapprochement with the United States. Debate continues to rage about the effectiveness of the "deterrent postures" of both countries and about whether they have increased or undermined the stability of the India-Pakistan relationship.[20] What is clear, however, is that the tests had significant results: India got the "breathing space" it needed to concentrate on economic development and on modernizing its conventional military; Pakistan succeeded in balancing India's conventional superiority and arguably in reducing the likelihood of another decisive defeat at the hands of its military.

Second, India's rise has brought about the forging of a "strategic partnership" with the United States, changing the dynamics not just of their relationship but also of American and Indian relations with Beijing. India and the United States are very far from being fully fledged allies, but they now cooperate in ways that were inconceivable before 1998—in intelligence sharing, for example, and counterterrorism. In turn, this new partnership has helped American-aligned states in the region to improve their relations with New Delhi, not least Australia and Japan, which have both moved to improve security cooperation and open economic opportunities in India. And although neither India nor the United States wants to try to "contain" China, both see value in coordinating their hedging strategies against the risk of more assertive (or even outright aggressive) Chinese policies that might emerge in the near future.

Third, India's rise is changing the nature and intensity of its relations with states across the Asia-Pacific, from West Asia to the Southwest Pacific. As they develop their economies, China and India are becoming more dependent on key commodities, above all oil and natural gas; on the security of the transit routes; and on the stability of their suppliers. Chinese and Indian attention is thus increasingly focused on the sea lanes of communication in the Indian Ocean—and on their control.[21]

Both China and India are increasingly bound to the Middle Eastern states; India is particularly dependent on Iranian crude oil. This growing dependence is beginning to have effects on both the security order in the Middle East and the wider Asia-Pacific, effects that will become more pronounced as the United States moves to energy independence. As China and India become ever more involved in the dynamics of the Middle East, driven by their interests in its hydrocarbon resources, they will reshape its security order. Arguably, they already have; Chinese support has helped to stymie Western efforts to aid Syria's rebels in its civil war, while Indian connections to Iran have led to exceptions being made to US and UN sanctions on Tehran concerning its oil exports.

India's impact on the regional security order has been diminished, however, by its elites' ingrained attitudes concerning the West, alliances, and multilateralism. India's elites have long been suspicious of Western motives, scarred by their experience of European imperialism. They remain wary, especially of American intentions in the region and therefore also of American initiatives. Nehru's intellectual legacy makes them cautious about realist tools like military alliances—redolent, as they are, of the kind of power politics he disdained—along with multilateralism, except when it is under United Nations auspices or driven by developing countries. India has now several "strategic partnerships" with

Asian states—not least with Indonesia, Japan, Singapore, and Vietnam—as well as a partnership, of sorts, with China. And India is a member of several important regional institutions, including the East Asia Summit. But India remains reluctant to make the kinds of binding commitments inherent in alliances or supranational institutions, and it remains unclear whether these strategic partnerships, which vary dramatically in content and momentum, will provide the kind of returns that India seeks in terms of economic investment, trade, and security.

Like India, Japan may have claims to great power status in the future. But as is discussed in the next section, it is unclear whether Japan aspires to this status and whether it is close to acquiring it.

Japan

Japan's rise to great power status in the late nineteenth and early twentieth centuries was unique. Facing increasing challenges from Western imperialism in Asia, Japan responded by replacing feudal isolation with a modern economic, political, and social system. The slogan of "strong army, rich nation" (*fukokukyōhei*) encapsulated the government's strategic thinking during this period; Japan would modernize by building up its economic and military structures to achieve greater prestige and power.[22]

Japan's strategy proved remarkably effective, and the country quickly sought to take a leading role in international affairs. Indeed, unlike much of Asia, Japan was not only able to resist Western incursions but also soon came to emulate Western imperialism.[23] Starting in 1876, when it became increasingly active on the Korean Peninsula, until its empire reached its zenith in 1942, Japan amassed colonies throughout Asia and the Pacific. By August 1945, however, Japan's strategic overreach and the increasing militarization of its foreign policies had led the country to disaster; in its push for empire, Japan had suffered immense physical, economic, and social destruction while also inflicting enormous suffering on the peoples of Asia and the Pacific.[24]

The legacies of these horrors, and the comprehensive reforms implemented by the United States–led Occupation authorities in Japan from 1945 until 1952, had a major impact on the country's postwar society. They also continue to resonate in Japan's strategic thinking today. The immediate Occupation reforms focused on democratization, demilitarization, and economic revitalization, with an important change being the introduction of a new Constitution. For foreign affairs, it was Article 9 of this new document that left the greatest legacy, underpinning the **pacifism** of Japan's postwar security strategy by renouncing "war as a sovereign right."[25] When the Occupation ended, Japan entered into an alliance with its former enemy, the United States, which would underpin Japan's national security strategy for the remainder of the Cold War and beyond. Yet the relationship with the United States was an unequal one. Japan's security was now provided by the United States, in return for the provision of bases for US military forces in Japan. In the ideological and strategic confrontation of the Cold War, Japan had become an important but junior ally of the United States within the San Francisco alliance system.[26]

This unequal relationship required Japan to reassess its strategic priorities. The country now focused on economic development while leaving its security to the United States. This policy was encouraged by the United States, but political leaders like former prime minister Yoshida Shigeru also resisted demands for the country to quickly rebuild its defense capabilities. Indeed, Yoshida and his successors were so successful in these endeavors that the country's grand strategy came to be known as the **Yoshida Doctrine**.[27] The subsequent story of Japan's Cold War strategic thinking is one of the nation's leadership coming to terms with the implications of these new, largely self-imposed constraints.

By the 1970s Japan was no longer a country devastated by war but was again a leading economic power. Yet it remained largely absent from the security field, and implemented a range of pacifist policies such as those based on its three nonnuclear principles—banning the possession, manufacture, and introduction of nuclear weapons into the country.[28] According to constructivists, this has been evidence that Japan's grand strategy has not followed a realist path but has instead been based on the structure of the state and a mix of social and legal norms. These norms have included the primacy of economics over security and a "postwar culture of antimilitarism." Others have continued to view Japan in realist terms, but as a "mercantile" rather than traditional realist.[29] Japan undoubtedly developed new ways of thinking about "security," broadening the term's meaning from the late 1970s, under the banner of **comprehensive security**, to also include energy, food, and other types of security.[30]

When the Cold War came to an end, however, the international environment underpinning Japan's strategic thinking disappeared. Japan was again searching for a new international role. The 1991 Gulf War exposed the nation to accusations of free-riding on the United States' security guarantee and of resorting to "checkbook **diplomacy**" as a result of its failure to do little more than provide financial support, albeit to the tune of $13 billion. The shifting post–Cold War order also prompted wider reassessments. Realists soon came to argue that the changes to the international order would increasingly push Japan toward accepting a more normal international role; to not do this would be a "structural anomaly." Such diplomatic humiliations and wider changes would lead the country to engage in a lengthy debate over its international role (table 3.3).[31]

Great Power Japan?

Given this complicated strategic legacy, is Japan a great power today? If measured by defense spending, the answer is yes. At $50.1 billion in 2013, Japan's spending on defense remains the seventh highest in the world, and Japan's defense forces

Table 3.3 Japan: Key Dates

Date	Event
1876	Began colonial expansion
August, 1, 1894–April, 17, 1895	First Sino-Japanese War
July 7, 1937–September 9, 1945	Second Sino-Japanese War
December 7, 1941	Entered World War II
September 2, 1945	Surrendered to end World War II
1945–52	Occupied by the United States
1947[a]	Adopted its Constitution

[a]The Constitution was promulgated in 1946, but it is generally referred to as "the 1947 Constitution" after the year it came into effect. Government of Japan, *The Constitution of Japan* (Tokyo: Government of Japan, 2015), http://japan.kantei.go.jp/constitution_and_government_of_japan/constitution_e.html.

have been described as "the most modern Asian armed forces in terms of equipment."[32] This is true despite the fact that the country spends less than 1 percent of its gross domestic product (GDP) on defense. Although its security strategy remains defensive, Japan continues to strengthen its hard power capabilities. In August 2013, for example, Japan launched the 27,000-ton helicopter carrier the *Izumo*, allowing it to deliver air defense more effectively over its distant territories. In its National Security Strategy of December 2013, Japan announced that it would build a "comprehensive defense architecture" while also focusing on the protection of the nation's territorial integrity.[33]

The question of whether Japan is a great power, therefore, has so far revolved largely around whether Japan desires to be one. Power, as constructivists might argue, is what states make of it.[34] In the debate that followed the 1991 Gulf War, politicians and commentators began to question whether the country needed to become a "**normal nation**" (*futsū no kuni*) once again. During this debate, figures such as Ichirō Ozawa argued that Japan should take up greater responsibilities as part of the international community.[35] This would include working more closely with the United Nations and other multilateral organizations in activities such as international peacekeeping and official development assistance. For others, however, becoming a normal nation entailed removing the postwar restrictions on national security, such as Article 9 of the Constitution and the Yoshida Doctrine. They argued that Japan needed to become a country more capable of defending itself and its partners. Since 2001 prime ministers such as Jun'ichirō Koizumi and Shinzō Abe have strongly argued that becoming a normal nation required changing the Constitution, relaxing restrictions on arms exports, and giving Japan the right to **collective self-defense**, that is, the right to help defend others if they are under attack.[36]

Japan's self-conception of its international role thus remains ambivalent.

Takashi Inoguchi and Paul Bacon have argued that during the 1990s, Japan first sought to define itself as a "global civilian power" before becoming a "global ordinary power."[37] Conversely, Yoshihide Soeya has argued that Japan's national strategy has effectively remained consistent—a "de facto "middle power" choice."[38] Soeya sees the idea of middle power diplomacy as an appropriate encapsulation of Yoshida's decision to rely on the United States for security and instead focus on economic development. This decision has been maintained into the post–Cold War period, despite the efforts of nationalist politicians such as Koizumi and Abe. Indeed, Japan's policies, Soeya argues, "hardly reflect the sort of diplomacy conducted by strategically independent great powers."[39]

Yet if Japan is a middle power, it is an atypical one. In addition to its high level of defense spending, as well as the exceptional technical proficiency of its defense forces, Japan has considerable capacity to upgrade its strategic capabilities. As the world's third-largest economy, and being still far more advanced technologically than other countries in the Asia-Pacific, it enjoys the industrial base to further develop its defense forces, even within current political constraints. Japan's capacity to influence regional diplomacy through links resulting from its extensive trade and official development assistance also continues to be strong, even if Japan does not dominate the region's economy as it once did. Japan remains a key ally of the United States. If it could further reorient its national security strategy, continue strengthening its diplomatic partnerships around the region, and press ahead with economic reform, Japan could once again become a strategically independent great power.[40]

Japan and the Regional Order: In Decline or Resurgent?

How might a shift in Japanese power—toward either middle power or great power status—affect the regional order? Until now, Japan's postwar grand strategy toward the region has been heavily reliant on strategic reassurance.[41] That is, mindful of the mistrust in the region due to its legacy of aggression from World War II, Japan accepted the limitations on its own power as a way to reassure the region about its intentions. The vital alliance with the United States provided a so-called cap in the bottle of potential Japanese militarism by sidestepping the need for Japan to rapidly rebuild its own defense capabilities. To this, Japan added an extraordinarily cautious defense policy known as defensive defense (*senshubōei*), while committing to never again play a "unilateral military role" in the region.[42] At the turn of the century, therefore, it could be argued that Japan was a "circumscribed balancer": weary of creating countervailing alliances and avoidant of developing offensive grand strategies.[43]

Japan's potential impact on the Asia-Pacific order, of course, presupposes an already set order in the region. Yet the region is undergoing dramatic change. Japan's strategic reassurance and circumscribed balancing followed from its dominance in Asia, the continuance of American hegemony in the region, and the recognition of these realities by China—all elements of the Cold War order. The rise of China and the emergence of clear Sino-Japanese rivalry since 2010, however, have raised doubts about all these assumptions. To understand how Japan might affect the regional order, therefore, it is necessary to consider simultaneously how these changes could reshape Japan's future strategic circumstances and the type of power it now wishes to be.

One can envisage a number of different scenarios that Japan might face in the coming decade or two and then extrapolate Japanese responses and likely effects from there. At present, a regional order shaped by Sino-American strategic rivalry seems most likely. Faced with such a strategic rivalry scenario, Japan would most likely seek to reinforce its alliance with the United States and expand its relationships with other US partners. In fact, signs of this scenario are already apparent, with Japan establishing a range of strategic partnerships around the region, most notably with Australia and India.[44] Continuity would be the main feature of such a scenario, and Japan's great power status would have the least impact on the regional order, because the pressures of a new balance of power across Asia would push Japan even further into the region's American-led security framework. These pressures would, however, also push Japan away from being a circumscribed balancer or constrained great power following a policy of strategic reassurance; indeed, Prime Minister Abe has already moved Japan toward a policy of collective self-defense. Under such a scenario, it can be assumed that this transition would continue and Japan would become a more assertive actor in international affairs, though not necessarily a great power.

Such circumstances might have a particularly destabilizing effect on regional order under an abandonment scenario—that is, if Japan were to be abandoned as part of a new balance of power in the region. If the United States were to retrench from Asia rather than consolidate its alliance system, and if this were to lead to Japan's loss of America's security guarantee, Japan could respond, as realists have long suggested, by rapidly building up its own military capabilities, such as a nuclear deterrence capability, to quickly become a traditional great power.[45] This would likely trigger a major regional arms race, heightening the risk of a confrontation between Japan and other regional powers, particularly China. A US departure from the region, however, appears a remote possibility in light of the Obama administration's efforts to boost the United States' engagement with the region and the strong possibility of Obama's successor following a similar, if not even more assertive, approach.[46]

Abandonment, however, may not come about as part of a US strategy of retrenchment. Under a **concert-of-powers** scenario—whereby the United States and China agree to a mutually constructed regional order—Japan would not be abandoned by the United States so much as sidelined as the United States came to a strategic accommodation with China. According to Hugh White, a Concert of Asia would require Japan to become a more independent actor, allowing a more even balance to emerge between the United States and China.[47] Yet it is not clear that Japan, sidelined by the other great powers, could then manage this kind of transition without upsetting the Sino-American accommodation. The question, then, is whether Japan would become a destabilizing force in this new order. To avoid such a scenario, Japan would need to strengthen its reassurance policy rather than discard it, perhaps by recommitting to its postwar antimilitarist norms. Yet the mixture of nationalism and security assertiveness pushed by the Abe administration belies such a possibility.

In the end, Japan's status will depend heavily on its own capacity to build and maintain the attributes of a great power. Japan has stagnated economically for more than twenty years now—the so-called lost decades—and has only rarely demonstrated the capacity or willingness to undertake the reforms necessary for recovery. If this situation persists, economic stagnation and demographic decline will make Japan's great power prospects considerably different than those of younger, more dynamic India. Optimistically, a slow decline implies that Japan would become a "marginal weight" in regional affairs, happy with its middle power status and continuing to support its alliance with the United States, but without being a central player.[48] Under a scenario where the United States retrenches from Asia, a declining Japan could choose to accommodate a rising China, or continue this accommodation according to some, and so choose the middle power path.[49] Conversely, extended decline could be dangerous if it were to lead to something like the abandonment scenario mentioned above. In decline, and in attempting to reverse such a middle power choice, Japan could become more nationalistic and a destabilizing force in the region.[50]

All these scenarios contain significant elements of risk for both Japan and Asia. An accommodation of China—particularly regarding the dispute over the Senkaku/Diaoyu Islands—would assist in defusing the growing Sino-American rivalry and potentially help create a concert situation like the one described above. This could be Japan's most positive contribution to a stable regional order. Yet accepting the resulting middle power status would be unpalatable to Japanese leaders. Greater nationalism in the political class, especially since Abe became prime minister in December 2012, reduces the likelihood that Japan will accept such a status.[51] Abe has asserted that Japan will not become a "tier two nation."[52] Instead, Japan's troubled relationship with China provides further impetus for it to become a great power once again, while also suggesting that any transition is unlikely to be smooth.

Conclusion

Neither India nor Japan is likely to emerge as a great power anytime soon, but both will play major roles in the regional security order. Conscious of China's rise and its own limited economic and military power, India, like many other regional states, has adopted a hedging strategy: concentrating as its first priority on raising its population out of poverty; acquiring nuclear weapons to buy it time to modernize its conventional forces; and tentatively engaging with selected strategic partners and regional institutions. This strategy is not likely to change for a decade or two, unless India suffers some dramatic endogenous or exogenous shock, such as a severe downturn in its economic fortunes or a surprise attack from China or Pakistan.

Japan's strategic direction, by comparison, seems poised for change. Certainly, its currently ambiguous status—not quite a great power, but much more than a middle power—complicates the regional security order and, in particular, Sino-American relations. Yet which strategic trajectory provides the best route to stability, either for the region or for Japan, is unclear. Certainly, Japan has sought to develop its relationships across the region as well as its own security capabilities, and its latent power means that it could extend its diplomatic and military capacities considerably. However, its troublesome relationship with China and its confused vision for its own international status means that Japan has become less able to provide strategic reassurance to the region than in the past.

In conclusion, it should be observed that the policies of both India and Japan call into question a number of key theoretical arguments. Neither has behaved as realists would expect: putting security first, building military power, and using force to secure their interests. Instead, India remains restrained, even when subjected to cross-border attacks and incursions from Pakistan or China, and focused on domestic challenges, while Japan continues to adhere to the postwar limitations placed upon its security policies, notwithstanding recent changes. Both nations continue to aspire to being normative or middle powers, influencing the world with ideas rather than coercion. But at the same time, democratic India does not always behave as liberals would predict—it remains wary about the supposed benefits of economic interdependence and multilateral security pacts. In both states, there are active debates about what sorts of powers India and Japan should be, lending some support to the constructivist argument that ideas and identities shape policies and behavior.

Key Points

- India is an emerging power, but it is some way from becoming a great power—even if its elites decided that it ought to be a conventional, instead of a "normative," great power.
- India remains very cautious about binding multilateral agreements or even bilateral alliances, preferring to pursue a hedging strategy to deal with the rise of China and the relative decline of the United States.
- The legacy of Japan's imperial past, wartime experience, and recovery during the Cold War continues to play a major role in Japan's contemporary strategic thinking. Of particular importance has been the Yoshida Doctrine.

- The factors shaping Japan's great power status are as much political and institutional as they are material. Yet Japan remains ambivalent about the kind of great power it wishes to become.
- A fundamental transformation of Japan's international status could destabilize the regional security order. Two key variables in this respect are Japan's relationship with the United States and the latter's own rivalry with China.

Questions

1. Is Indian foreign policy still driven more by ideology than self-interest?

2. Did India's acquisition of nuclear weapons in 1998 enhance or diminish its security?

3. How has the Yoshida Doctrine shaped Japan's strategic thinking and international status?

4. What does it mean for Japan to become a "normal nation"?

5. Are India and Japan great powers or not great powers? How do the different security study theories help to answer this question?

Guide to Further Reading

On India

Cohen, Stephen P. *India: Emerging Power.* Washington, DC: Brookings Institution Press, 2004.
A classic—and very accessible—American assessment of India's foreign and security policy.

Ganguly, Šumit, ed. *India's Foreign Policy: Retrospect and Prospect.* New Delhi: Oxford University Press, 2012.

Mohan Raja, C. *Crossing the Rubicon: The Shaping of India's New Foreign Policy.* New Delhi: Viking, 2003.
Another classic work, but written by a prominent Indian realist, arguing that India's foreign policy has been transformed since 1991.

Panagariya, Arvind. *India: The Emerging Giant.* New York: Oxford University Press, 2008.
The major study of India's changing economy.

On Japan

Goh, Evelyn. "How Japan Matters in the Evolving East Asia Security Order." *International Affairs* 87, no. 4 (2011): 887–902. http://dx.doi.org/10.1111/j.1468-2346.2011.01009.x.

Pyle, Kenneth B. *Japan Rising: The Resurgence of Japanese Power and Purpose.* New York: Public Affairs, 2007.
Written by one of the foremost historians of Japan, Japan Rising traces the country's engagement with the world from the nineteenth century until today.

Samuels, Richard J. *Securing Japan: Tokyo's Grand Strategy and the Future of East Asia.* Ithaca, NY: Cornell University Press, 2007.
: *Samuels's Securing Japan remains the seminal work on Japanese strategic thinking in the twentieth century.*

Taylor, Brendan. "Asia's Century and the Problem of Japan's Centrality." *International Affairs* 87, no. 4 (2011): 871–85. http://dx.doi.org/10.1111/j.1468-2346.2011.01008.x.
: *In this important debate, Goh and Taylor lay out contrasting prognoses of Japan's future significance to the Asia-Pacific order.*

Notes

1. Jonathan Holslag, *China and India: Prospects for Peace* (New York: Columbia University Press, 2009), 10–11.

2. Stephen P. Cohen, *India: Emerging Power* (Washington, DC: Brookings Institution Press, 2004), 41–43.

3. Christophe Jaffrelot, "India's Look East Policy: An Asianist Strategy in Perspective," *India Review* 2, no. 3 (2003): 35–68.

4. Ian Hall, "The Engagement of India," in *The Engagement of India: Strategies and Responses*, ed. Ian Hall (Washington, DC: Georgetown University Press, 2014).

5. George J. Gilboy and Eric Heginbotham, *Chinese and Indian Strategic Behavior: Growing Power and Alarm* (Cambridge: Cambridge University Press, 2012), 254.

6. Harsh V. Pant, "The US-India Nuclear Pact: Policy, Process, and Great Power Politics," *Asian Security* 5, no. 3 (2009): 273–95.

7. Ian Hall, "China Crisis? Indian Strategy, Political Realism and the Chinese Challenge," *Asian Security* 8, no. 1 (2012): 84–92.

8. Sunil Khilnani and Srinath Raghavan, *Nonalignment 2.0: A Foreign & Strategic Policy for India in the 21st Century* (Harmondsworth, UK: Penguin, 2013).

9. For an interesting discussion, see Amrita Narlikar, "All That Glitters Is Not Gold: India's Rise to Power," *Third World Quarterly* 28, no. 5 (2007): 983–96; and Cohen, *India: Emerging Power.*

10. Ian Hall, "Normative Power India?" in *China, India, and International Society*, ed. Jamie Gaskarth (Lanham, MD: Rowman & Littlefield, 2015). On Nehru's views, see inter alia Ritu Sharma, "Nehru's World-View: An Alternative to the Superpowers' Model of International Relations," *India Quarterly: A Journal of International Affairs* 45, no. 4 (1989): 324–32. The classic study of the idea of a "normative power" is by Ian Manners, "Normative Power Europe: A Contradiction in Terms?" *Journal of Common Market Studies* 40, no. 2 (2002): 235–58.

11. Ian Hall, "The Other Exception? India as a Rising Power," *Australian Journal of International Affairs* 64, no. 5 (2010): 601–11.

12. See, e.g., Raja C. Mohan, *Crossing the Rubicon: The Shaping of India's New Foreign Policy* (New Delhi: Viking, 2003).

13. Martin Wight, *Power Politics*, ed. Hedley Bull and Carsten Holbraad (London: Leicester University Press, 1995), 50.

14. World Bank, "Gross Domestic Product, 2012," http://databank.worldbank.org/data/download/GDP.pdf.

15. For the best general discussion of India's changing economy, see Arvind Panagariya, *India: The Emerging Giant* (New York: Oxford University Press, 2008).

16. Stephen P. Cohen and Sunil Dasgupta, *Arming without Aiming: India's Military Modernization* (Washington, DC: Brookings Institution Press, 2012).

17. On this topic, see Brahma Chellaney, *Water: Asia's New Battleground* (Washington, DC: Georgetown University Press, 2013).

18. See, especially, Daniel Markey, "Developing India's Foreign Policy 'Software,'" *Asia Policy* 8, no. 1 (2009): 73–96; and Manjari Chatterjee Miller, "India's Feeble Foreign Policy," *Foreign Affairs*, May–June 2013.

19. Ian Hall, "The Requirements of Nuclear Stability in South Asia," *Nonproliferation Review* 21, nos. 3–4 (2014): 355–71.

20. See, e.g., Šumit Ganguly and S. Paul Kapur, *India, Pakistan, and the Bomb: Debating Nuclear Stability in South Asia* (New York: Columbia University Press, 2010).

21. See C. Raja Mohan, *Samudra Manthan: Sino-Indian Rivalry in the Indo-Pacific* (Washington, DC: Carnegie Endowment for International Peace, 2012).

22. Richard J. Samuels, *Securing Japan: Tokyo's Grand Strategy and the Future of East Asia* (Ithaca, NY: Cornell University Press, 2007), 15–18. See also Richard J. Samuels, *"Rich Nation, Strong Army": National Security and the Technological Transformation of Japan* (Ithaca, NY: Cornell University Press, 1994), 34–42.

23. Akira Iriye, *Japan and the Wider World: From the Mid-Nineteenth Century to the Present* (London: Longman, 1997), 11–22.

24. See Kenneth B. Pyle, *Japan Rising: The Resurgence of Japanese Power and Purpose* (New York: Public Affairs, 2007), 185–96. For a study of the suffering inflicted by Japan on the region, see Yoshimi Yoshiaki, *Comfort Women: Sexual Slavery in the Japanese Military during World War II*, trans. Suzanne O'Brien (New York: Columbia University Press, 1995). On the strategic thinking that led Japan to war, see Jack Snyder, *Myths of Empire: Domestic Politics and International Ambition* (Ithaca, NY: Cornell University Press, 1991), 112–52.

25. Prime Minister of Japan and His Cabinet, *Constitution of Japan*, May 3, 1947, www .kantei.go.jp/foreign/constitution_and_government_of_japan/constitution_e.html. Concerning the drafting of the postwar Constitution, see John W. Dower, *Embracing Defeat: Japan in the Wake of World War Two* (New York: W. W. Norton), 346–404.

26. See Roger Buckley, *US-Japan Alliance Diplomacy, 1945–1990* (Cambridge: Cambridge University Press, 1992), 27–114; and Kent E. Calder, "Securing Security through Prosperity: The San Francisco System in Comparative Perspective," *Pacific Review* 17, no. 1 (2004): 135–57. See also Michael J. Green, "Balance of Power," in *US-Japan Relations in a Changing World*, ed. Steven K. Vogel (Washington, DC: Brookings Institution Press, 2002), 9–34, at 12–14.

27. Samuels, *Securing Japan*, 29–37.

28. On Satō's three nonnuclear principles, see Andrew L. Oros, *Normalizing Japan: Politics, Identity, and the Evolution of Security Practice* (Stanford, CA: Stanford University Press, 2008), 102–7.

29. For Japan's postwar culture of antimilitarism, see Thomas U. Berger, "From Sword to Chrysanthemum: Japan's Culture of Anti-Militarism," *International Security* 17, no. 4 (1993): 119–50. On the wider norms underpinning Japanese security policy during this period, see Peter J. Katzenstein and Nobuo Okawara, "Japan's National Security: Structures,

Norms, and Policies," *International Security* 17, no. 4 (1993): 84–118. For Japan's mercantile realism, see Eric Heginbotham and Richard J. Samuels, "Mercantile Realism and Japanese Foreign Policy," *International Security* 22, no. 4 (1998): 171–203. See also H. D. P. Envall, *Japanese Diplomacy: The Role of Leadership* (Albany: State University of New York Press, 2015), 63–67.

30. Eiichi Katahara, "Japan's Concept of Comprehensive Security in the Post–Cold War World," in *Power and Prosperity: Economics and Security Linkages in the Asia-Pacific*, ed. Susan L. Shirk and Christopher P. Twomey (New Brunswick, NJ: Transaction, 1996), 213–31, at 214–19.

31. Christopher W. Hughes, *Japan's Security Agenda: Military, Economic, and Environmental Dimensions* (Boulder, CO: Lynne Rienner, 2004), 159–60. Also see Samuels, *Securing Japan*, 66–67. For the 1990s' structural realist viewpoint on Japan as a "structural anomaly," see Kenneth N. Waltz, "The Emerging Structure of International Politics," *International Security* 18, no. 2 (1993): 44–79.

32. International Institute for Strategic Studies, "Asia," *Military Balance* 114, no. 1 (2014): 250. See also International Institute for Strategic Studies, "Comparative Defense Statistics," *Military Balance* 114, no. 1 (2014): 23.

33. Regarding the *Izumo*, see Benjamin Schreer, "The 'Aircraft Carrier' That Isn't: Japan's New Helicopter Destroyer," *Strategist*, August 19, 2013, http://www.aspistrategist .org.au/the-aircraft-carrier-that-isnt-japans-new-helicopter-destroyer/. On Japan's comprehensive security architecture, see Government of Japan, "National Security Strategy," December 17, 2013, 15.

34. Alexander Wendt, "Anarchy Is What States Make of It: The Social Construction of Power Politics," *International Organization* 46, no. 2 (1992): 391–425.

35. David A. Welch, "Embracing Normalcy: Toward a Japanese 'National Strategy,'" in *Japan as a "Normal Country"? A Nation in Search of Its Place in the World*, ed. Yoshihide Soeya, Masayuki Tadokoro, and David A. Welch (Toronto: University of Toronto Press, 2011), 16–37, at 29–30.

36. H. D. P. Envall, "Transforming Security Politics: Koizumi Jun'ichiro and the Gaullist Tradition in Japan," *Electronic Journal of Contemporary Japanese Studies* 8, no. 2 (2008), www .japanesestudies.org.uk/articles/2008/Envall.html; H. D. P. Envall, "Japan's Strategic Challenges in Asia," *AIIA Policy Commentary* 15 (2013): 37–43, at 38–39.

37. Takashi Inoguchi and Paul Bacon, "Japan's Emerging Role as a 'Global Ordinary Power,'" *International Relations of the Asia-Pacific* 6, no. 1 (2006): 1–21, at 4–5.

38. Yoshihide Soeya, "A 'Normal' Middle Power: Interpreting Changes in Japanese Security Policy in the 1990s and After," in *Japan as a "Normal Country?"* ed. Soeya, Tadokoro, and Welch, 72–97, at 74.

39. Yoshihide Soeya, "Japanese Middle-Power Diplomacy," East Asia Forum, November 27, 2012, http://www.eastasiaforum.org/2012/11/22/japanese-middle-power -diplomacy/. Also see Yoshihide Soeya, *Nihon no MidoruPawāGaikō: Sengo Nihon no Sentaku to Kōsō* (Japan's "Middle Power Diplomacy": The Choices and Plans of Postwar Japan) (Tokyo: Chikuma Shinsho, 2005), 23–24.

40. Michael J. Green, "Japan Is Back: Unbundling Japan's Grand Strategy," *Analysis*, Lowy Institute for International Policy, December 2013, 3–4. On recent movements in Japan's approach to arms exports, see "Abe Set to Lift Arms Embargo to Nations Involved in

Armed Conflicts," *Asahi Shimbun*, February 23, 2014, ajw.asahi.com/article/behind_news/politics/AJ201402230016.

41. Paul Midford, "The Logic of Reassurance and Japan's Grand Strategy," *Security Studies* 11, no. 3 (2002): 1–43, at 1.

42. Ibid., 28–30. Regarding Japan's role in the regionalization process, see Evelyn Goh, "How Japan Matters in the Evolving East Asia Security Order," *International Affairs* 87, no. 4 (2011): 887–902, at 899–900.

43. Christopher P. Twomey, "Japan, a Circumscribed Balancer: Building on Defensive Realism to Make Predictions about East Asian Security," *Security Studies* 9, no. 4 (2000): 167–205, at 168.

44. Thomas S. Wilkins, "Japan's Alliance Diversification: A Comparative Analysis of the Indian and Australian Strategic Partnerships," *International Relations of the Asia-Pacific* 11, no. 1 (2011): 115–55; H. D. P. Envall, "Japan's India Engagement: From Different Worlds to Strategic Partners," in *Engagement of India*, ed. Hall, 39–59.

45. Samuels, *Securing Japan*, 151.

46. Mark E. Manyin, Stephen Daggett, Ben Dolven, Susan V. Lawrence, Michael F. Martin, Ronald O'Rourke, and Bruce Vaughn, *Pivot to the Pacific? The Obama Administration's "Rebalancing" toward Asia*, CRS Report for Congress R42448 (Washington, DC: Congressional Research Service, 2012).

47. Hugh White, *The China Choice: Why America Should Share Power* (Collingwood, Australia: Black Inc., 2012), 142–43.

48. On Japan becoming a "marginal weight," see Brendan Taylor, "Asia's Century and the Problem of Japan's Centrality," *International Affairs* 87, no. 4 (2011): 871–85, at 882.

49. Björn Jerdén and Linus Hagström, "Rethinking Japan's China Policy: Japan as an Accommodator in the Rise of China, 1978–2011," *Journal of East Asian Studies* 12, no. 2 (2012): 215–50.

50. H. D. P. Envall, "Implications for Asia in Japan's Economic Decline," *East Asia Forum Quarterly* 2, no. 3 (2010): 28–29, at 28.

51. H. D. P. Envall, "Abe's Yasukuni Visit Escalates Tensions in Asia," *East Asia Forum*, January 3, 2014, www.eastasiaforum.org/2014/01/03/abes-yasukuni-visit-escalates-tensions-in-asia/.

52. Cited by Green, "Japan Is Back," 2.

4 Are the Middle Powers on a Collision Course in the Asia-Pacific?

Andrew Carr

Reader's Guide

The "middle powers" of the Asia-Pacific—states with some capacity for self-defense and multilateral influence—are increasingly asserting a voice on the key security challenges of our time. They are also the focus of much of the United States' diplomatic activity following its "pivot" or "rebalance" to the region. Not all their influence is benign. While these states have fewer resources and influence than the great powers, they can still cause conflict and pose risks for the region's continued peace and prosperity. This chapter looks at the region's middle powers and examines how they affect the regional security order, both on their own and working with a great power ally, for good and ill.

Introduction

Continuing the hierarchical organization of states delineated in the first part of this book, this chapter looks at the medium-sized countries. It argues (as does chapter 5) that a thorough understanding of Asia-Pacific security requires not just looking at the great powers—such as the United States, the People's Republic of China (hereafter, China), India, and Japan—but also at the region's middle-sized and even small states. These states may not attract as many newspaper headlines, but they are crucial states to consider if the region is to avoid war and build peace and prosperity.

Scholars currently debate how to identify a **middle power**, and how to delineate what influence they have and in what circumstances. Yet there seems a general agreement that below the level of the great powers, but separate from the small states, are a range of middle power states. These states play a wide variety of roles in the security order of the Asia-Pacific. They are the supporters and followers that support the status quo established by the great powers. This occurs via alliances and partnerships as well as through **multilateral** forums. They can act as niche leaders advocating change or as regional smoke detectors, ringing the alarm bell when they detect changes occurring to the regional structure that they see as hazardous. Finally, they can be responsible for initiating conflict, either with other mid-sized and smaller states or potentially even against the largest states of the region. This chapter outlines how scholars identify middle powers, discusses which states in the region are middle powers, and looks at why they might be the "builders" or "breakers" of peace and prosperity in the region during this century.

The major security theories have tended to downplay the role of non–great power states. However, though not devoting much attention to these mid-range and smaller states, each of the major theories has different ideas about how these states might achieve influence. Realists—who tend to be the most dismissive of non–great powers being influential— would point to the need for military capacity, and weapons that might overcome weakness such as weapons of mass destruction or using asymmetric tactics (see chapter 8, which discusses how nonstate actors achieve results against larger states). Liberals, meanwhile, stress how smaller states, through **coalition** building and multilateral forums, might be able to build collective weight and gain greater international legitimacy. Finally, constructivists demonstrate how middle power states can try to overcome their material weakness with ideational strength—that is, how these powers are often seen as activists and norm entrepreneurs, trying to promote new ideas about behavior, which in turn might support their interests.

What Is a Middle Power?

While many prime ministers and presidents refer to their own countries as middle powers (including Australia, Indonesia, Mexico, South Korea, and Turkey), scholars have had great difficulty determining what exactly is meant by the label "middle power." These states are not great powers; they fall short of the test established in chapter 1 for great power status—global influence and strength in war—but they may have some say in how their own region operates (e.g., influencing the institutions, laws, norms, and customs). This separates these states from the smallest nations, which tend to ignore international affairs, focus on their own corner of the world, and often have little if any capacity for influencing the regional and international order.

So what are the middle powers, beyond what they are not—neither large nor small? This is not a trivial question. If we know what a middle power country is, we can know not only what can rightly be considered one (thus telling us about which countries are likely to be influential in the future) but we may also have a better idea about how they might go about having influence. A good definition should have some predictive capacity. It should tell us something about what we can expect these states to be able to do. If middle powers tend to act in a particular way (e.g., focusing on multilateral institutions and coalition building), then we should expect to see this occurring in the region. If middle powers adopt a realist approach and tend to try to balance against rising powers, or bandwagon with the largest existing power, then that too tells us something about how the Asia-Pacific's security future might play out. Good definitions help scholars separate the noise of global politics (the many stories, reports, speeches, and videos of leaders and the region's daily changes) and find the details that matter, the telling shifts in policy, and the important new claims or ideas.

This book uses a "systemic impact" definition to identify middle powers as *states that can protect their core interests and can initiate or lead a change in a specific aspect of the existing international order.*[1] This definition is focused on the behavior of middle powers, but many scholars have also used other definitions based on the material resources that influence state power such as population, economic size, and military capacity. One of the most common approaches has been economic rankings, using **gross domestic product** (GDP), to rank states. Under this approach, the great powers, the United States and China, are respectively the 1st- and 2nd-largest economies in the world, with Japan and India not far behind. These states are an order of magnitude larger than their neighbors in the Asia-Pacific. The middle powers are identified as those states that fall into the second and third groups of nations, typically ranked around 8th to 20th in the size of their economies, though some scholars use 10th to 30th as the best range.[2] It is thus important to recognize that though middle power states might be found in between the largest and smallest states, they are not explicitly "in the middle" of the 193 countries recognized by the United Nations.

The advantage of ranking states with quantitative measurements is that it provides a clear-cut order. But the risk is that these measurements may not be that useful for understanding state power or behavior. Knowing whether one state is wealthier than another does not tell us where or how well they can apply this resource to achieve their national interests. Although most scholars accept that economic wealth translates into a state's power, it is not always a direct relationship. Economic wealth can be used to purchase more military equipment, support allies with assistance, or apply economic pressure (e.g., sanctions). However, most economic wealth within a state, especially in capitalist economies, is possessed by private individuals and companies. As is discussed at greater length in chapter 5, quantitative definitions have a useful role, but they are less effective when seeking to understand the behavior and impact of mid-sized and smaller states.

Given these concerns about measuring power, scholars in the mid-1990s sought a "behavioral" approach to defining middle powers, seeking to understand these states by what they do rather than what they have.[3] This literature has tended to focus on diplomatic influence through multilateral forums, and this chapter takes that as a starting point. Thus, there is a view that by gathering together with other similarly sized states—or by using the rules, norms, and laws of institutions to their advantage—"what middle powers may lack in economic, political, or military clout, they can often make up with quick and thoughtful diplomatic footwork."[4] Scholars have demonstrated that middle powers are able to secure regional or global influence on important issues by working through multilateral forums.[5] This literature recognizes liberal and constructivist insights that state power and influence are not just the sum of material assets, as many realists assume.

Critics of this approach have argued that the desire to engage with multilateral forums, especially global forums, is not common to all middle-sized states.[6] While Canada and Australia have played important roles globally, far more common are states such as Turkey and Indonesia, which concentrate on regional diplomatic influence. In addition, the behavioral approach often downplays the significance of material and military issues for these states, sometimes presenting an undue impression that they are more moral or just than other types of states.[7] In contrast, however, scholars in the security studies field have long noted that the middle powers play important roles as supportive military allies or may seek to influence regional balances of power or use their militaries to secure their national interests.[8] Therefore, it is important to include both diplomatic and military areas to understand potential middle power influence. Following this literature, this chapter judges that to be a middle power, alongside diplomatic strength, a state should be able to militarily defend itself. It may not be able to win every single battle, especially against great powers, but it should be able to make it a difficult and costly experience for another state to attempt to do so. This also means that we should expect middle powers to have some viable military capacity to use against smaller states, though not to seek out war or be confident of victory in the same way that great powers might be.

To properly explore the role of middle powers in the security order of the Asia-Pacific, this book has sought a definition that offers insight into both the diplomatic and military behavior of middle powers. As this chapter shows, though the middle powers may present themselves as "builders" using diplomatic means to bring peace, they can also be "breakers" risking conflict in the region and failing to cooperate in ways that could have a significant influence on the prospect of a great power war (table 4.1).

Which Countries Are Middle Powers?

Having established our definition, which countries in the Asia-Pacific meet the criteria to be middle powers? Using the "systemic impact" definition, eight countries in the region can be classified as middle powers: Australia, Canada, Indonesia, North Korea, Pakistan, South Korea, Russia, and Vietnam. On one level, this is a very diverse group of states. Indonesia has a population of 247 million, and

Pakistan has 180 million, while Australia has just 23 million. In terms of economic size, Russia is the 8th-largest economy, Canada and Australia are 11th and 12th, and South Korea and Indonesia are just behind them, at 15th and 16th. However, Pakistan and Vietnam are far down this list, at, respectively, 43rd and 55th. But as noted above, such ranking numbers are not everything. Both Vietnam and Pakistan have a strong regional position. Vietnam, in various forms, managed to resist the use of force by the French, Americans, and Chinese during the Cold War. Today it is central to the emerging dispute in the South China Sea and the long-term direction and significance of **the Association of Southeast Asian Nations (ASEAN).** Pakistan is a nuclear power that has already played a central role in the "War on Terror," withstanding intense pressure both from the United States and from those seeking to carve up the country to form an Islamic homeland. In the long term, Pakistan's relationship with India and China will be crucial for the stability of South Asia and the Indian Ocean.

Debates about which country is or is not a middle power tend to focus on those states at the upper and lower levels of the category. Russia would strike many as being a great power. Certainly this is a rank it has had in the past and to which it might like to return, but it is not clear that it can do so. While it has the world's largest nuclear arsenal, Russia suffers from a corrupt authoritarian regime that has become highly centralized around the presidential office, currently held by Vladimir Putin. It is experiencing a deep decline in its population—by 2025 its population is expected to be 20 million lower than it was when the Cold War ended. As a result, the Russian economy is struggling, with the state increasingly dependent upon the current price of oil and natural gas exports. When these factors change, so does the capacity of the state, which makes Russia's power unreliable and vulnerable. At the same time, when the prices of Russia's exports are high, this can empower it, giving an uneven quality to its foreign policy. Russia has proven willing to break international norms in recent years, most notably with its annexation of Crimea and its border war with Ukraine. Though the former was a deft military move, the latter has demonstrated many of the weaknesses of modern Russia, while furthering the international isolation and economic stagnation of this former power.

According to traditional quantitative definitions, North Korea would not be viewed as a middle power. It is one of the world's economically weakest nations, ranking 119th overall. Its population ranks 50th globally, half the size of its neighbor South Korea. North Korea also wields no persuasive global influence. If **soft power**, the so-called attractiveness and appeal of a nation, could be measured, North Korea would get a zero. No one outside its borders looks to the "hermit kingdom" to provide leadership or even assistance during crises. Yet North Korea does have one clear advantage: a willingness to break all international norms, laws, and standards to protect the hereditary regime that rules the country. This has included the successful pursuit of nuclear weapons. In doing so, North Korea

Table 4.1 Middle Powers: Key Facts

Characteristic	Australia[a]	Canada[b]	Indonesia[c]	North Korea[d]	Pakistan[e]	Russia[f]	South Korea[g]	Vietnam[h]
Official languages	No official language, English main language	English, French	Bahasa Indonesia	Korean	Urdu, English	Russian	Korean	Vietnamese
Population	23.6 million	35.5 million	251.5 million	24.9 million (2013)	186.3 million	143.7 million	50.4 million	90.6 million
Land area (square kilometers)	7,690,000	9,971,000	1,905,000	121,000	796,000	17,098,000	99,000	332,000
GDP (dollars)	1,440.0 billion	1,788.7 billion	888.6 billion	15.5 billion (2013)	250.1 billion	1,857.5 billion	1,416.9 billion	186.0 billion
Principal export destination	China (33.9% of merchandise exports)	United States (76.8%)	Japan (13.1%)	China (72.0%)	United States (14.7%)	Netherlands (13.7%)	China (25.4%)	United States (18.1%)
Principal industry[i]	Mining	Transportation equipment	Petroleum and natural gas	Military products	Textiles and apparel	Mining and extractive industries	Electronics	Food processing
Number of military personnel[j]	56,750	65,700	676,500	1,379,000	947,800	1,260,000	659,500	522,000

Note: All figures for 2014, except where marked otherwise and for military personnel, the figures for which are all for 2013.

[a] Australian Department of Foreign Affairs and Trade, *Australia Country Fact Sheet* (Canberra: Commonwealth of Australia, 2014), http://dfat.gov.au/trade/resources/Documents/aust.pdf.

[b] Australian Department of Foreign Affairs and Trade, *Canada Country Fact Sheet* (Canberra: Commonwealth of Australia, 2014), http://dfat.gov.au/trade/resources/Documents/can.pdf.

[c] Australian Department of Foreign Affairs and Trade, *Indonesia Country Fact Sheet* (Canberra: Commonwealth of Australia, 2014), http://dfat.gov.au/trade/resources/Documents/indo.pdf.

[d] Australian Department of Foreign Affairs and Trade, *DPRK Country Fact Sheet* (Canberra: Commonwealth of Australia, 2014), http://dfat.gov.au/trade/resources/Documents/dprk.pdf.

[e] Australian Department of Foreign Affairs and Trade, *Pakistan Country Fact Sheet* (Canberra: Commonwealth of Australia, 2014), http://dfat.gov.au/trade/resources/Documents/paki.pdf.

[f] Australian Department of Foreign Affairs and Trade, *Russia Country Fact Sheet* (Canberra: Commonwealth of Australia, 2014), http://dfat.gov.au/trade/resources/Documents/russ.pdf.

[g] Australian Department of Foreign Affairs and Trade, *Republic of Korea Country Fact Sheet* (Canberra: Commonwealth of Australia, 2014), http://dfat.gov.au/trade/resources/Documents/rkor.pdf.

[h] Australian Department of Foreign Affairs and Trade, *Vietnam Country Fact Sheet* (Canberra: Commonwealth of Australia, 2014), http://dfat.gov.au/trade/resources/Documents/viet.pdf.

[i] Central Intelligence Agency, *Field Listing: Industries, CIA World Factbook 2014* (Washington, DC: Central Intelligence Agency, 2014), https://cia.gov/library/publications/the-world-factbook/fields/2090.html.

[j] World Bank, *Armed Forces Personnel, Total* (Washington, DC: World Bank, 2015), http://data.worldbank.org/indicator/MS.MIL.TOTL.P1.

withstood pressure from five important neighbors, South Korea, China, Russia, Japan, and the United States. Despite being the smallest and economically weakest country in Northeast Asia, surrounded on all sides and militarily outclassed, North Korea has an outsized influence on the regional environment. Pyongyang has achieved major agreements, gaining support and aid for its embattled country, has defied pressure to disarm, and has regularly acted in a provocative and threatening manner without reprisal. Certainly the brutal way North Korea is ruled has had a great human cost, but this small nation nonetheless achieves middle power status, in regularly warding off potential great power coercion and achieving internationally significant **diplomacy** (albeit just for its own limited self-interests). For all these reasons, many scholars and policymakers fear that the likeliest cause of a future conflict in East Asia will be action by the North Koreans.[9] This would be a case of history repeating itself, with North Korea having started the Korean War in 1950, one of the worst conflicts of the half-century-long Cold War. Yet, as we will see in the pages that follow, there are many other middle powers, often with much more positive international records, that also have the potential to cause regional conflicts.

Somewhere in the middle of the middle powers are Australia, Canada, Indonesia, and South Korea. These states all face questions about their capacity, given internal dynamics, but each of them already has a proven track record. Canada, like the United States, has made the choice to participate in the Asia-Pacific. It does not need to do so, and there is a significant debate within its defense establishment on whether it should instead focus on Europe. Yet, in light of the economic opportunities in the region and the US rebalance to the region, Canada has sought to become much more active in East Asia. Doubts have also been raised about Australia's willingness to be a middle power, though it has the financial basis as the 12th-largest economy in the world (as of 2016) to be one, if it chooses. It also has a clear record of diplomatic significance, often seeming to be the model middle power.[10] Less doubt applies to Indonesia and South Korea. By virtue of their important geographic positions and increasing displays of leadership (explored below), these states have a clear claim to middle power status. One final caveat, however, remains: Scholars have long noticed a "paradox of unrealized power" for middle powers, that is, "seeming to wield less overall power than their position relative to the major powers, . . . [though] the degree of underachievement of potential obviously differs widely."[11] Unlike the great or small nations, there is a degree of choice for middle powers. Some achieve significant influence without readily having the capacity; some clearly have the capacity but never try to apply it. Intention matters greatly, as does the skill with which individual leaders guide their nations.[12] To answer this chapter's central question, whether the middle powers are on a collision course in Asia, this chapter now explores two key types of behavior that are common to middle powers: They can act as builders or as breakers. What roles the middle powers of the Asia-Pacific will choose to play will go a long way toward deciding whether there is peace or conflict in the region in the coming years.

Builders

The best-known role of middle powers is as the builders and standard bearers for international order. As states weaker than the great powers, middle powers look to institutions, international law, norms, and other social forces that can help them to achieve their own national interests, prevent challenges to themselves or their key partners, and reduce the risk of great power conflict within their region. For liberals and constructivists, this is a promising avenue, whereas realists tend to be much more pessimistic about the ultimate impact of this behavior. This role, one that is often championed by the middle powers as a prime justification for their importance, can be understood as the role of builder.

The assumption of most security studies scholars (and international relations in general) is that the tactics and strategy of states are determined by their relative power. While liberals and constructivists argue that **democratic** and **authoritarian** states may seek different goals, there is much to the realist argument that the means of influence and the behavior of states is often determined by their relative power and influence. For example, though the United States may today be a great power, and therefore be able to operate in a **unilateral** fashion, it acted quite differently during its early years, when it was a weak, small, or middle power state. In the first decades after the American Revolution, the United States often sought the security of cooperation and alliances with larger states (switching between Great Britain and France), and it advocated on behalf of international law and justice as the foundations of international behavior.[13] As the United States grew more powerful, its goals of national security and encouraging the spread of liberal democratic government remained, but its tactics changed to more coercive, confrontational means. The amount of power a state has does not explain everything about its behavior, but can help explain certain aspects.

As we have seen above, middle powers can expect to have some level of power and influence—not enough to change things entirely their way, but certainly enough to shape key issues. One popular method they use is combining their weight via coalitions with other like-minded states or using international force multipliers, such as institutions or appeals to international law and norms. In the Asia-Pacific, there are many examples of middle power states undertaking building roles to encourage the establishment, maintenance, or adjustment of the international order in a way that ensures their security and protects their values. So common is this behavior that one approach to defining middle powers has emerged that entails focusing on the tendency of these states to act in coalitions or via institutions.[14] Among the region's middle powers, North Korea, Pakistan, and Vietnam have historically shown the least interest in or capacity for this type of endeavor. In time, however, it is certainly possible that we will see Pakistan and Vietnam considering how to build coalitions to shape the outcomes of issues such as the future of Afghanistan and the South China Sea.

The states most scholars would associate with the building role are Australia and Canada. Both share the same origins as the United States (as British colonies), and therefore they viewed much of the post–World War II regional security order built by the United States in the Asia-Pacific as very supportive of their own agendas. In turn, they have acted as defenders of this security order, while at the same time advancing ideas that they think are in line with the order's general principles and ideals.[15] Through their close connections to the United States, both countries have not only tried to work with other middle powers but have also tried to encourage Washington to focus on certain areas they felt were missing. For example, Canada has advocated banning land mines through the Convention on the Prohibition of the Use, Stockpiling, Production, and Transfer of Anti-Personnel Mines and on Their Destruction. They have also attempted to establish forums that support their particular interests as well as those of their great power ally, the United States, such as Australia's leadership to build the Asia-Pacific Economic Cooperation forum (APEC). When the middle powers are committed and have a good justification for their proposed change, they can sometimes be very successful in achieving it. Thus the region's economic order owes much to the activism of middle power builders such as Australia and Canada.[16]

While there is a clear record, some doubt the merits and capacity of these Western middle powers as builders in the so-called Asian Century. In the twenty-first century, the momentum and enthusiasm for middle power building roles has now passed to states such as South Korea, Indonesia, and Russia. All three have been strongly focused on encouraging multilateral institutions and approaches to the problems of the Asia-Pacific. The state acting closest to the traditional model of middle powers is South Korea. During the last five to ten years, South Korea has established itself as an activist state with a focus on issues such as environmental governance—including addressing climate change—and nuclear nonproliferation.[17] Seoul has also played an important role in the Six-Party Talks with Pyongyang to try and persuade it to give up its nuclear weapons program. Though this effort has been paused since 2009 (and may well have ceased to exist), the Six-Party Talks were an example of the **minilateralist** approach, whereby a small group of relevant states cooperate to address a common issue. When middle powers can participate in these small forums, it offers a real opportunity for them to wield significant influence and make an important contribution to regional stability and security.

Indonesia has also acted as a significant power within its own region. In the 1960s Jakarta was instrumental in the establishment of ASEAN. Since then, it has consistently sought to ensure that multilateralism, via the ASEAN framework, remains Southeast Asia's primary method of regional economic and security organization. ASEAN gives the region's middle and small powers a much more powerful say than they would be able to achieve alone. This can be seen in the case of the disputed islands in the South China Sea. Many scholars believe that China's best strategy to achieve its claims is by trying to divide ASEAN. But if it must negotiate with this full coalition of middle and smaller states, it will have a much harder time claiming the full land area it seeks. Notably, Beijing appears unwilling to directly challenge Indonesia in the way it has challenged the Philippines and Vietnam, and Jakarta's leadership will be crucial to ASEAN's future influence.

During the past decade, as Indonesia has moved from being an authoritarian regime to a well-functioning democracy, it has begun to create a role for itself as an advocate for issues such as democracy in Asia. In this way, though it is not an ally of the United States, it has still helped to build and incorporate Southeast Asia into the United States–established regional order. Sometimes this has meant working with the United States (e.g., against the emergence of terrorist threats after 2001), and sometimes it has meant challenging the United States' views—for instance, downplaying the focus on human rights and encouraging a stronger respect for national **sovereignty**. Depending on one's perspective, some of these changes may not be positive, but they certainly constitute an effort by the middle powers to build, expand, and sustain the existing regional order. Inevitably, regional orders require changes to their nature and functioning to ensure that all states can peacefully participate within them. Therefore, the middle powers—as regional leaders, like Indonesia; as **idealists**, like Canada; or as mediators, like Australia and South Korea—can play an important role in adjusting how the regional order operates in order to ensure the continuation of the existing framework.

Russia has also undertaken a building role, though its purpose has been not to sustain but to dismantle the United States–led regional security order. Not everything Russia does should be attributed to anti-Americanism, but it distrusts many of the institutions and frameworks established by the United States and its allies. Russia believes and argues that replacing the existing order with a new approach will not only better serve its own interests but will also support the ideals and values of most countries in Asia. Therefore, Russia has established the Eurasian Forum, which involves Central Asian nations (an area not covered within this book's focus on the Asia-Pacific); has joined and supported China's Central Asia–focused Shanghai Cooperation Organization; and has contributed to push regional forums such as APEC in its preferred direction. Russia has also expanded its bilateral ties with South Korea, Japan, and Vietnam. Just as the United States is pivoting toward the Asia-Pacific, in part because of the opportunity to build trade links and foster economic growth, Russia is also looking east and seeking greater economic integration with the region.

Over the next few decades, it is likely that the middle powers' building role will become increasingly important. It has long been recognized that the regional security order established by the United States after World War II is under increasing pressure to change and adjust to recognize the "rise of the rest"—including new technologies and new types of state interaction.[18] Whether the middle powers work to support this order (as Australia, Canada, and South Korea in particular do), or whether they turn against it and embrace an alternative vision (as Russia does, and as Indonesia, Vietnam, and others could do), will be vital for determining the region's long-term security.

If scholarly predictions of a **multipolar** order in the Asia-Pacific—based on the assumption that several great powers (the United States, China, Japan, and India) are competing for influence and authority—are true, the middle powers will have more capacity for influence and significance. In such an environment, these mid-sized states, with both diplomatic and military capacity, will become key coalition partners. While many of the possible solutions to the regional challenges will first need to be discussed between the great powers (as is explained in chapters 1 and 2 and in the conclusion), for any "regional" agreement to be viable it will need the middle powers' consent if not active support. If they withhold this, their role could soon turn from being builders of the regional order to breakers of it.

Breakers

Most scholarship on middle powers has examined the positive contribution that these powers make to the established order. This is certainly encouraged by the middle powers themselves, which have even created labels for their behavior, such as "good international citizenship."[19] Yet this is far from the only type of behavior that one may see these states undertaking. It is also possible, indeed probable, that the middle powers will act as breakers of the security of the Asia-Pacific.

A state acts as a breaker if it pursues its national interests ahead of both the maintenance of the regional order and the interests of the region and the great powers. In a number of cases, small states and middle powers have sought to obtain an advantage or opportunity with full knowledge that their behavior will significantly increase the risk of conflict. The same growing significance and capacity for influence that gives the middle powers a greater say in the importance of multilateralism in the Asia-Pacific can also enable them to cause havoc and damage in specific circumstances. There are several types of breaker actions that one can identify from middle powers: They can undermine the actions of other states that are acting as builders; they can fail to cooperate; they can make high demands on their great power allies; and they can even start conflicts.

An inevitable tension when some middle powers are acting as builders is that other states in the region will not agree with the vision, ideas, or proposed membership and will seek to undermine what is occurring. While one should not expect the middle powers to always get along or have similar interests, it is notable that a common reason middle powers are not as effective on the world stage is because of undermining by other middle powers (or small states, at the right time and place). Nations like Australia and Canada have often suffered from this, as outsiders attempting to influence a region where they are culturally divergent. In 1989 and 1993, when Australia was seeking to create APEC and later increase its regional status, the Malaysian prime minister Mahathir Mohamad strongly opposed the Australian initiatives on the ground that Australia's status as part of the Asia-Pacific was questionable.[20] In 2008 another Australian initiative, this time the Asia-Pacific Community (APC), was killed off by another small state in the region, Singapore. The APC concept was badly launched, but its vision of a multilateral forum to bring China and the United States together to discuss key regional security questions was important.[21] After the APC idea was torpedoed by Singapore and other states to ensure the supremacy of ASEAN, an ASEAN-led forum, the **East Asia Summit**, was later re-formed to incorporate the central idea behind the APC. This ensured that ASEAN, and not Australia, got the credit and stayed in the influential chair's position.

ASEAN, however, also has its own internal struggles. As described in chapter 11, though many middle powers and small states see multilateralism as vital to their capacity to manage the great powers in the Asia-Pacific, they also regularly undermine the initiatives of other smaller states in order to achieve their specific **national interests** or to curry favor with the great powers. As noted above, one of the most significant regional challenges is that of maritime security and the South China Sea. And though the middle powers and small states have different claims (sometimes over the same territory), they also have a united interest in forcing China to negotiate via multilateral forums, where the middle powers and small states can assert the most influence and seek to offset the weight and power of Beijing. Yet in 2012, when it looked like ASEAN was beginning to establish this principle, Cambodia, at the behest of China, broke ranks and undermined the ASEAN forum's work.[22] Meanwhile, there have been concerns that Vietnam and especially the Philippines are not working through ASEAN but are seeking their own initiatives, which are making multilateral resolution with China harder to achieve.

Along with failing to cooperate despite mutual interests in multilateral arenas, middle powers may also act as breakers and not cooperate when dealing with issues on which they otherwise should, such as transnational security issues. Some challenges—such as piracy, terrorism, irregular migration, and drug smuggling—are widely acknowledged as issues that are beyond the capacity of any one state, given they involve nonstate actors that move easily between borders and may have sophisticated resources. As such, it is imperative to cooperate to deal with these issues. Yet too often, disputes between the middle powers have undermined or prevented the establishment of practical regional responses to these issues. This was evident in the hunt in early 2014 for a Malaysian airplane, MH370, which had disappeared mid-flight en route from Malaysia to China. Because the plane's transponder had been turned off, it fell to the region's countries to share their intelligence and radar data to help searchers locate the missing plane. Yet the sensitivity of the middle powers—like Australia, Singapore, Vietnam, and Indonesia—to release what they knew (due to worries that other countries could learn about the capacity of their surveillance systems) caused several major delays in the eventual tracking of the plane.[23] While it is unlikely that anything could have been done in the first few days to save those on board, faster sharing of intelligence would have perhaps made recovery of the plane easier and certainly cheaper.

Another way in which the middle powers can act as breakers that harm the regional peace is by making high demands on their great power allies and partners. While the great powers obviously have the capacity to say no to an ally's request, there is a widespread fear in capitals such as Washington that if they do so, it will lead other states to question the larger states' credibility and thus their own relationships with the larger state. Currently, the United States has military alliances with Japan, South Korea, the Philippines, Thailand, and Australia; it also has growing ties with several other nations in the region, such as Taiwan, Singapore, and Indonesia. If Washington ignores calls for help or action by any one of these countries, it risks a chain reaction of concern and doubt in other capitals in the region.[24] China, India, and Japan do not currently have the same challenge of managing alliances with smaller states, but they may well confront similar situations in the future, given their efforts to build and strengthen ties around the region as part of their great power competition. The risk here is that if a middle power or small state were to become embroiled in a conflict, it could place their great power ally in the position of risking war with a large state or risking the great power's entire alliance structure. This has happened before. In 1914 a dispute between a middle power state, Serbia, and the crumbling empire of Austria-Hungary led both sides to demand support from their great power allies (Russia and Germany, respectively). This in turn set off "the war to end all wars" in the region. While the great powers could have (and perhaps, in retrospect, should have) said no to their allies, they found themselves in a lose–lose position because of the actions of more aggressive middle and smaller states.[25]

The same pattern has occurred in the Asia-Pacific. In 1950 North Korea sought the permission of the USSR and China to launch an invasion of South Korea. This led to a United Nations resolution condemning the invasion and a UN-sanctioned mission to repulse the attack. This brought the United States onto the Korean Peninsula and quickly into conflict with China. This was one of a number of proxy wars in the region, whereby conflicts between and within the middle powers and smaller states drew in the great powers as part of their struggle for supremacy in the Cold War. The middle powers knew this struggle made their allegiance all the more important, and in turn they exploited this for their own benefit. Taiwan currently trades on its democratic and capitalist identity to ensure that it gains the support and protection of the United States. A Taiwanese declaration of independence—something it openly debated in the mid-1990s, and an action it could still decide to take—is seen as another extremely likely cause of great power conflict between the United States and China. Currently, the longest-running border dispute—with regular violent clashes—is between two smaller powers, Thailand and Cambodia. Likewise, Malaysia and the Philippines have found themselves nearly coming to blows over disputed claims for eastern Sabah. Just because a country is smaller does not mean that its people are any less nationalistic, or that its leaders are any more inclined to seek peaceful resolutions when they think conflict or at least confrontation can achieve their interests. Indeed, Russia seems to be treating its declining power and economic capacity as an invitation to throw its weight around, attacking Georgia in 2008 and Ukraine in 2014, along with making regular threats against Europe and the United States.

As you read the rest of this book, try to look not only at how the flashpoints and challenges may involve the great powers like the United States and China but also at what role a middle power or small state could take in changing the outcome. Of the major flashpoints that scholars fear in Asia—such as the Korean Peninsula, the Taiwan Strait, the Malacca Strait, the South China Sea, and the East China Sea—only the last of these does not directly feature middle power states as central and independent actors. North Korea, in particular, could (and at times seems determined to) provoke a conflict with South Korea and/or Japan. In 2012 and 2014 the North Koreans launched shells into South Korean territory (with South Korea responding in kind). In both cases, an escalation was avoided; but many scholars see the Korean Peninsula as a highly volatile situation where massive casualties and costs could occur.[26] If either of the great powers was drawn in (e.g., the United States seeking to protect South Korea from a future North Korean attack), it would be difficult to see how the other great power could resist acting. In Southeast Asia, the likelihood of conflict is lower, but there is still a chance that it might occur. Significant tension remains between Indonesia, Singapore, and Malaysia over control of the vital Malacca Strait, through which 20 percent of the world's seaborne oil trade passes.[27] Meanwhile, China is far from the only country trying to obtain access to the potential resources of the South China Sea. And though Beijing could live without the potential riches of the South China Sea, for smaller countries like the Philippines and Vietnam, maintaining access to the resources in the region is a vital concern for which they have already demonstrated a willingness to fight.

Conclusion

This book has begun its exploration of Asia-Pacific security by identifying the hierarchy of states. Hierarchy seems to be the most accurate way to describe the region—as opposed to unipolar or multipolar.[28] There are clear and important differences in the capacities of the great power states, such as the United States and China, from nearby states such as Japan and India. Equally, these states stand alone when compared with the rest of the region. But as this chapter demonstrates, a comprehensive understanding of the Asia-Pacific today cannot just end with the major powers. Middle power states are more numerous and just as active—for good or ill—as their larger partners. In the following chapter, a similar argument is made on behalf of the role of the region's small states and microstates. Not all states matter at all times, but in the right circumstances middle powers and even small states can be either builders or breakers of the regional order.

Builders contribute to the establishment of regional institutions that may form the basis for solving regional security challenges (e.g., terrorism and piracy) or avoiding great power conflict, as well as developing norms and conventions that can help encourage diplomacy and negotiation instead of armed conflict. Breakers undermine the initiatives of others, undertake risky actions themselves, or place their great power allies in difficult positions. While Washington and Beijing will have the final say over whether the Asia-Pacific remains peaceful or turns violent, the flashpoints where they may need to choose between war or peace, and the forums in which the great powers negotiate regional order and prosperity, will often be the products of middle power activism. No analysis of security in the region can therefore ignore the roles of these mid-sized states.

Key Points

- While there are many ways to identify middle powers, each with different strengths, this chapter argues that these are *states that can protect their core interests and can initiate or lead a change in a specific aspect of the existing international order.*
- Eight states in Asia fit this pattern; they are (in alphabetical order): Australia, Canada, Indonesia, North Korea, Pakistan, South Korea, Russia, and Vietnam.
- Middle powers may act as builders, helping to establish regional institutions and norms that provide conflict resolution options and help address regional security challenges.
- Middle powers may also act as breakers, undermining the actions of other states, dragging their great power allies into difficult situations, or even launching conflicts of their own.

Questions

1. Is there something unique about middle power states?

2. Can middle powers act as "good international citizens," as some have argued? If so, what effect do you think this can have on regional security challenges?

3. Why do middle powers tend to prefer working in coalitions and institutions?

4. Why do great powers struggle to control the actions of middle power states?

5. Which middle powers do you think will be the most important for regional security in the Asia-Pacific?

Guide to Further Reading

Ba, Alice D. *Renegotiating East and Southeast Asia: Region, Regionalism, and the Association of Southeast Asian Nations.* Stanford, CA: Stanford University Press, 2009.
  A very detailed but highly readable description of the evolution of ASEAN and multilateralism in Southeast Asia.

Cooper, Andrew F., Richard A. Higgott, and Kim Richard Nossal. *Relocating Middle Powers: Australia and Canada in a Changing World Order.* Melbourne: Melbourne University Press, 1993.
  The classic text on middle powers as builders in the global order. Focuses on Australia and Canada in the 1980s and early 1990s.

Chapnick, Adam. "The Canadian Middle Power Myth." *International Journal* (Toronto) 55, no. 2 (2000): 188–206. http://dx.doi.org/10.2307/40203476.
  It is important to remain skeptical about the claims made by policymakers over the influence their state possesses. This classic article demonstrates how identifying as a middle power served the interests of Canadian leaders, but did not always reflect the reality of the nation's power.

Dibb, Paul. "Towards a New Balance of Power in Asia." *Adelphi Series* 35, 1995.
  Though it was published nearly twenty years ago, this article is one of the few to comprehensively detail how the middle powers are seeking to develop their military capacity and the way technological change could have an important role on their future security and influence.

Goh, Evelyn. *The Struggle for Order: Hegemony, Hierarchy, and Transition in Post-Cold War East Asia.* New York: Oxford University Press, 2013. http://dx.doi.org/10.1093/acprof :oso/9780199599363.001.0001.
  This book shows how the United States has sought to negotiate a new role in Asia in the period since the Cold War. It shows how this negotiation has gone between the United States and the middle and smaller powers.

Ikenberry, G. John, and Mo Jongryn. *The Rise of Korean Leadership: Emerging Powers and Liberal International Order.* New York: Palgrave Macmillan, 2013. http://dx.doi.org /10.1057/9781137351128.
  South Korea is one of the most important and active middle powers in Asia today. This book explores how it has embraced the idea of middle power leadership in fields such as nuclear nonproliferation and mitigating climate change.

Notes

1. For a full explanation of this approach, see Andrew Carr, "Is Australia a Middle Power? A Systemic Impact Approach," *Australian Journal of International Affairs* 68 (2014): 70–84.

2. See Bruce Gilley and Andrew O'Neill, eds., *Middle Powers and the Rise of China* (Washington, DC: Georgetown University Press, 2014), 5.

3. Andrew F. Cooper, Richard A. Higgott, and Kim Richard Nossal, *Relocating Middle Powers: Australia and Canada in a Changing World Order* (Melbourne: Melbourne University Press, 1993).

4. Gareth Evans and Bruce Grant, *Australia's Foreign Relations: In the World of the 1990s* (Carlton: Melbourne University Press, 1995), 347.

5. Andrew Cooper, ed., *Niche Diplomacy: Middle Powers after the Cold War* (New York: St. Martin's Press, 1997); Andrew Carr, *Winning the Peace: Australia's Campaign to Change the Asia-Pacific* (Melbourne: Melbourne University Press, 2015); David Cooper, "Challenging Contemporary Notions of Middle Power Influence: Implications of the Proliferation Security Initiative for 'Middle Power Theory,'" *Foreign Policy Analysis* 7, no. 3 (2011): 317–36.

6. Eduard Jordaan, "The Concept of a Middle Power in International Relations: Distinguishing between Emerging and Traditional Middle Powers," *Politikon: South African Journal of Political Studies* 30 (2003): 168.

7. E.g., see Bernard Wood, *The Middle Powers and the General Interest* (Ottawa: North-South Institute, 1988), 20.

8. See Carsten Holbraad, *Middle Powers in International Politics / Carsten Holbraad* (London: Macmillan Press, 1984), 12–13; Paul Dibb, "Towards a New Balance of Power in Asia," *Adelphi Series* 35, no. 295 (1995), 58; Richard Hill, *Maritime Strategy for Medium Powers* (Sydney: Croom Helm, 1986); Andrew Cooper, Richard Higgott, and Kim Richard Nossal, "Bound to Follow? Leadership and Followership in the Gulf Conflict," *Political Science Quarterly* 106, no. 3 (1991): 391–410.

9. Taw-Hwan Kwak and Seung-Ho Joo, *North Korea and Security Cooperation in Northeast Asia* (Farnham, UK: Ashgate, 2014), 5.

10 For an extended discussion of Australia's claim to middle power status, see Carr, *Winning the Peace.*

11. Wood, *Middle Powers*, 19.

12. Cooper, *Niche Diplomacy.*

13. Robert Kagan, *Dangerous Nation: America's Place in the World, from Its Earliest Days to the Dawn of the 20th Century* (New York: Knopf Doubleday, 2006).

14. Cooper, Higgott, and Nossal, *Relocating Middle Powers.*

15. Jordaan, "Concept."

16. Cooper, Higgott, and Nossal, *Relocating Middle Powers.*

17. G. John Ikenberry and Mo Jongryn, *The Rise of Korean Leadership: Emerging Powers and Liberal International Order* (New York: Palgrave Macmillan, 2013).

18. Charles A. Kupchan, "After Pax Americana: Benign Power, Regional Integration, and the Sources of a Stable Multipolarity," *International Security* 23 (1998): 73.

19. Alison Pert, *Australia as a Good International Citizen* (Sydney: Federation Press, 2014), 1.

20. Carol Johnson, Pal Ahluwalia, and Greg McCarthy, "Australia's Ambivalent Re-Imagining of Asia," *Australian Journal of Political Science* 45 (2010): 65.

21. Andrew Carr and Chris Roberts, "Foreign Policy," in *The Rudd Government: Australian Commonwealth Administration 2007-2010*, ed. Chris Aulich and Mark Evans (Canberra: ANU E-Press, 2010), 248, available at http://press.anu.edu.au?p=6031.

22. Bill Hayton, *The South China Sea: The Struggle for Power in Asia* (New Haven, CT: Yale University Press) 2014, 199.

23. Adam Taylor, "The Geopolitics of Asia Are Complicated, and So Is the Search for MH370," *Washington Post*, March 19, 2014, http://www.washingtonpost.com/blogs /worldviews/wp/2014/03/19/the-geopolitics-of-asia-are-complicated-and-so-is-the -search-for-mh370/.

24. Jeffrey Bader, Kenneth Lieberthal, and Michael McDevitt, *Keeping the South China Sea in Perspective*, Foreign Policy Brief (Washington, DC: Brookings Institution, 2014), http://www.brookings.edu/~/media/research/files/papers/2014/08/south-china-sea -perspective-bader-lieberthal-mcdevitt/south-china-sea-perspective-bader-lieberthal -mcdevitt.pdf.

25. Christopher Clark, *The Sleepwalkers: How Europe Went to War in 1914* (London: Penguin Books, 2012), 45.

26. Brendan Taylor, "The South China Sea Is Not a Flashpoint," *Washington Quarterly* 37 (2014): 101.

27. US Energy Information Administration, *The South China Sea Is an Important World Energy Trade Route* (Washington, DC: US Department of Energy, 2013), http://www.eia.gov /todayinenergy/detail.cfm?id=10671.

28. Evelyn Goh, *The Struggle for Order: Hegemony, Hierarchy, and Transition in Post–Cold War East Asia* (New York: Oxford University Press, 2013).

5 Why Are Small States a Security Concern in the Asia-Pacific?

Joanne Wallis

Reader's Guide

While security studies has traditionally focused on "great" and "middle" powers, this chapter examines the concept of "**small states**." It begins by considering the contested definition of a "small state," with popular measures based on the size of a state's population, economic indicators, and military capacity. The chapter then proposes a definition of a small state and identifies the small states in the Asia-Pacific. It also identifies a subcategory of "**microstates**," which tend to experience particular vulnerabilities that differentiate them from other small states. The chapter concludes by applying the security studies theories discussed in the book's introduction to small states in order to consider how they pursue their security. Applying these theories to small states can tell us interesting things about how states use their relative capabilities to undertake action and to influence other states in the region.

Introduction

As you have read in the introduction, traditional approaches to security studies focus on the roles of states and of power in the international system. While this book has so far focused on "great" and "middle" powers, this chapter considers the role of "small" states, and the subcategory of "microstates," in the Asia-Pacific.

Security studies often focuses on larger powers because it assumes that states with power will inevitably use that power and it is interested in the consequences. However, in the Asia-Pacific, the changing security order often means that great and middle powers are constrained in their use of power, whereas small states may have more freedom to act independently. This means that small states might be able to influence the outcomes of competition for influence between great powers and exploit power asymmetries to their advantage. It also means that the ability of great or middle powers to use deterrence or coercion to influence the behavior of small states might be reduced, thus undermining the power of those larger states.

In addition, under international law all states are considered sovereign and equal, regardless of their size or their power. This means that small states may sometimes have as much say in international and regional institutions—such as the **United Nations** and the **ASEAN Regional Forum**—as do great powers like the United States and the People's Republic of China (hereafter, China). As illustrated below in this chapter, this means that strong small states can shape the international and regional security agenda, for example, actions by South Pacific microstates to combat climate change. However, as is also discussed below, weak and small states can also pose a disproportionate security threat to larger states, not because they themselves can take military action against larger states but because they can act as havens for terrorists and transnational criminals that can target larger states.

Depending on the definition adopted, in the Asia-Pacific there are as many as thirty-one small states, consisting of fourteen small states and seventeen microstates. Small states and microstates represent the largest groups of states in the region, which suggests that these states are "simply too numerous and . . . too important to ignore."[1] Moreover, given that self-determination and secession movements are active in parts of the region, it is likely that new small states will emerge, and that they will increasingly need to be taken into account in studies of security issues in the region.

What Is a "Small State"?

The academic literature has struggled to reach a consensus concerning the definition of a "small state." Small states are often "defined by what they are not"; that is, small states are not great or middle powers.[2]

It is possible to measure smallness. The most popular measure is based on the size of a state's population because the data are easily available and comprehensible.[3] This is the measure used by international development institutions such as the **Commonwealth Secretariat**, the **UN Industrial Development Organization**, and the **World Bank**. If population size is used, it is necessary to set a threshold. At the higher end, an upper limit of between 10 million and 15 million people has been proposed.[4] In Europe states have generally been considered small if their population is lower than that of the Netherlands—currently 16.7 million[5]—although this threshold has recently been revised up to a population of 40 million.[6] Other scholars have used upper limits of 1 million,[7] 3 million,[8] and 5 million.[9] In their widely cited report, the Commonwealth Secretariat and World Bank Joint Task Force on Small States used an upper limit of just 1.5 million but also included larger states that share many of the same characteristics, such as Papua New Guinea (which has a population of 7.1 million).[10] Some authors distinguish between small states and microstates, which are sometimes defined as those that have a population either less than 1.5 million or less than 100,000.[11]

In the Asia-Pacific context, if a population of 40 million is taken as the upper measure of small statehood, there are thirty-one small states, ranging from Canada, with a population of 34.9 million, to Niue, with a population of 1,398. However, using a strict population measure results in including states in the small state category that are considered middle powers—such as Australia, Canada, and North Korea—as discussed in the previous chapter. For example, although Australia has a small population (23 million), it has a very large territory (7.6 million square kilometers) and economy (the twelfth-largest in the world). A strict population measure also excludes states such as Bangladesh (150 million), Burma/Myanmar (50.5 million), the Philippines (91.9 million), and Thailand (67.7 million), which although large in terms of population, have relatively small economies, comparatively minimal military power, and limited international influence, which might mean that they are better classified as small states. This suggests that though the population measure might constitute a satisfactory definition of a microstate, it is much more problematic when trying to identify small states (table 5.1).

Given that population is not always an accurate reflection of the size of a state's power, other authors measure the size of a state's territory or its economy.[12] These scholars are concerned with the relative power rather than population size of states, as a larger territory or developed economy can increase a state's power.

Table 5.1 Small States by Population, 2014

State	Population
Bangladesh	159.0 million
The Philippines	99.1 million
Thailand	67.7 million
Burma/Myanmar	53.4 million
Afghanistan	31.6 million
Malaysia	29.9 million
Nepal	28.2 million
Sri Lanka	20.6 million
Cambodia	15.3 million
Papua New Guinea	7.5 million
Laos	6.7 million
Singapore	5.5 million
New Zealand	4.5 million
Mongolia	2.9 million

Source: World Bank, "Population, Total," 2015, http://data.worldbank.org/indicator/SP.POP.TOTL.

They focus on states' military power,[13] or they emphasize their capabilities[14]—that is, their capacity and resources to act independently, without interference from, or dependence on, other states.[15] For example, although North Korea has a comparatively small population (24 million), it has significant military power—including nuclear weapons—and it has been able to act relatively independently of other states, even its close partner, China. This is well illustrated by its refusal to comply with international pressure to dispose of its nuclear weapons. Although North Korea could be classified as a small state according to its population, if other factors are taken into account, it is better considered a middle power.

This last point highlights the fact that there is a difference between a state being small and a state being weak. As Singapore illustrates, a state might have a small population (5.3 million), but it might also have strong political institutions, a highly developed economy (the 35th biggest in the world), and significant international or regional influence. The ability of small states such as Singapore to exercise this influence has been enhanced by globalization, which has reduced international barriers and increased the free flow of people, money, and ideas. This has allowed Singapore and certain other small states to capitalize on their location close to major markets or their technical expertise in order to develop significant international economic influence. Moreover, states with large populations, such as the Philippines, are not necessarily strong or able to exercise great power or influence at the regional or international level. Consequently, some authors add a psychological dimension to the definition of a small state, whereby a state is small if its "leaders consider that it can never, acting alone or in a small group, make a significant impact on the system."[16] In this way, small states can be distinguished from middle powers, which, as discussed in the previous chapter, consider that they are able to (at least occasionally) exert significant influence in the region.

To overcome these challenges and contradictions, this book adopts a definition of small states that takes account of a range of factors that indicate a state's capabilities and its capacity to act independently. It is proposed that a small state is one that *has fewer economic, military, and societal resources than great and middle powers and that consequently is vulnerable to interference by larger powers and finds it difficult to act independently.*

Adopting this definition, it is possible to classify the following fourteen Asia-Pacific states as small states: Afghanistan, Bangladesh, Cambodia, Laos, Malaysia, Mongolia, Burma/Myanmar, Nepal, New Zealand, Papua New Guinea, the Philippines, Singapore, Sri Lanka, and Thailand (table 5.2).

What Is a Microstate?

If a population of 1.5 million or less is used to identify microstates, there are seventeen in the Asia-Pacific, ranging from Niue to Timor-Leste, with a population

Table 5.2 Microstates by Population, 2014

State	Population
Timor-Leste	1,200,000
Fiji	886,000
Bhutan	765,000
Solomon Islands	572,000
Brunei	417,000
Maldives	401,000
Vanuatu	259,000
Samoa	192,000
Kiribati	110,000
Federated States of Micronesia	104,000
Tonga	106,000
Marshall Islands	53,000
Palau	21,000
Cook Islands	13,000
Tuvalu	10,000
Nauru	10,500
Niue	1,500

Source: World Bank, "Population, Total," 2015, http://data.worldbank.org/indicator/SP.POP.TOTL.

of 1.2 million. If a maximum population of 100,000 is used, there are six, which are all located in the South Pacific. Because there are several similarities between states in the region that have a population at or under the threshold of 1.5 million, in this book this threshold is used to define microstates. With the exceptions of Bhutan and Brunei, all the region's microstates are islands (or parts of islands), and of the islands, all but the Maldives are located in the South Pacific.

Many microstates experience particular physical, economic, social, and political vulnerabilities, which make them "especially susceptible to harm."[17] This vulnerability is said to be a consequence of the interaction of two sets of factors: first, the "incidence and intensity of risk and threat"; and second, the "ability to withstand risks and threats (resistance) and to 'bounce back' from their consequences (resilience)."[18] For example, island states in the South Pacific are particularly at risk of natural disasters and of climatic and ecological threats, as they are located near the volatile eastern edge of the Australasian tectonic plate and in the tropical belt of the large Pacific Ocean. However, their ability to withstand these risks is often limited, as they tend to have few economic resources and inadequate technical capacity to adopt strategies to protect themselves from, or adapt to, these threats. Papua New Guinea, though also located in the South Pacific and facing similar vulnerabilities, does not qualify as a microstate because its population exceeds the 1.5 million threshold. Despite this, Papua New Guinea shares many similarities with the South Pacific microstates, which might warrant it being considered part of this category. It should be noted that though microstates "remain vulnerable, they do not remain helpless."[19] Microstates can adopt policies to try to minimize their vulnerabilities and to capitalize on their opportunities. For example, Vanuatu and Fiji have capitalized on their pristine beaches and tropical weather to develop thriving tourism industries, which have in turn delivered significant resources to these states that they have used to enhance their resilience.[20]

Adopting the population definition of 1.5 million to identify microstates, the following seventeen Asia-Pacific nations are microstates: Bhutan, Brunei, the Cook Islands, the Federated States of Micronesia, Fiji, Kiribati, the Maldives, the Marshall Islands, Nauru, Niue, Palau, Samoa, the Solomon Islands, Timor-Leste, Tonga, Tuvalu, and Vanuatu.

What Do Security Studies Theories Tell Us About the Security Challenges Faced by Small States?

Applying security studies theories to small states can tell us how states use their relative capabilities to undertake action and to influence other states. **Realism** tells us that states with strong military and economic capabilities will use their power to pursue their interests. As a result, realists predict that the **security dilemma** will be particularly acute for small states because, compared with great and middle powers, they lack military and economic capabilities and are therefore likely to feel more threatened when larger states pursue their interests. This also means that the stakes are higher for small states when they deal with great and middle powers, as their comparative weakness means that they cannot afford to be cheated and that they are likely to experience severe consequences if they make mistakes or miscalculations.[21] These fears are evident in the Asia-Pacific, where China's rising military and economic power has left the small states close to its borders feeling vulnerable. As is discussed in chapter 7, China has made increasingly assertive claims to maritime territory in the South China Sea. In November 2013 China declared an "Air Defense Identification Zone" over the parts of the South China Sea that it claims are its territory, which requires aircraft from other states that enter the identification zone to identify themselves to the Chinese Ministry of Foreign Affairs or the Civil Aviation Administration of China and to follow any instructions given to them. Small states that have competing territorial claims to the sea, or whose aircraft frequently enter the area—such as the Philippines, Malaysia, and Brunei—are likely to feel threatened by China's moves and have expressed their concern about them. The stakes are high for these small states, for if they miscalculate their actions, they may face a severe Chinese response.

The **neorealist** concepts of bandwagoning and balancing also help to explain the behavior of Asia-Pacific small states. Neorealists predict that small states are more likely to bandwagon with a great power than to balance against it, to avoid immediate attack.[22] For example, in the Asia-Pacific, the Philippines' decision to renew its military ties with the United States, as well as Singapore's increasing closeness to the United States, may be evidence of small states attempting to bandwagon on the United States' great power status and to minimize the likelihood that they will be attacked by other states. In this regard, neorealists argue that hegemonic systems can provide a significant degree of security for small states, because the hegemon can prevent regional conflicts that would threaten the small state's security.[23] These situations are known as "patron–client" relationships, and involve the hegemon providing a security guarantee to the small state, often in return for other services, such as hosting military bases.[24]

In the Asia-Pacific, there has been a hegemonic system since the Korean War concluded in 1953, because the United States has been the only great power and has provided a security guarantee for the region, which helped prevent the outbreak of interstate war. Moreover, the United States' San Francisco alliance system provides an example of middle powers and small states bandwagoning with a great power, as discussed in chapter 1. After World War II, to counter any rival power, the United States sought to develop a series of alliances in the region, known as the "San Francisco System." This name derives from the San Francisco Peace Treaty process that occurred in September 1951, when forty-nine nations met at the San Francisco Opera House to conclude a treaty with Japan. Washington used this treaty-making process as a focal point, around which it created a regional defense network.

The San Francisco System is based on bilateral alliances and relationships. The United States' strongest defense treaty is with Japan, for after World War II the United States was keen to ensure that Japan did not fall into the Soviet orbit. Thus the United States sought to provide Japan with sufficient economic opportunity to act as a growth engine for the Asia-Pacific region, and it wanted to ensure that it could gain access to substantial bases in Japan. Because Australia, New Zealand, and the Philippines were extremely apprehensive over the prospect of revived Japanese aggression, the United States entered into defense treaties with them. In subsequent years, these treaties were supplemented by a defense treaty with Thailand and defense pacts with South Korea and Taiwan.

The San Francisco System is often described as a "hub-and-spokes" network, with the United States as the hub projecting its power into the region by way of these bilateral alliances and agreements. The system has a highly asymmetrical structure, particularly in military terms, as the United States possesses significant military power in comparison with the middle powers and small states with which it has relationships. When combined with the United States' vast national resources, the San Francisco System has allowed Washington to exert unparalleled hegemonic influence in the region since the conclusion of the Korean War. In exchange it has provided its partner middle powers and small states with a security guarantee.

However, neorealists caution that in a hegemonic system, small states will be constrained by the sole great power; if they bandwagon with that power, they will be reliant on it. Neorealists argue that in a competitive (balance-of-power) system, small states will have more room to maneuver, as they will be able to play competing great powers off against each other.[25] This was evident during the Cold War, when the region's small states became desirable allies to the United States and the Soviet Union in their competition for regional and global influence. This gave small states the ability to extract development assistance and other benefits from their patrons.[26] However, this leverage can be fleeting; thus, after the Cold War many small states found that these benefits ceased (or were at least diminished) because the balance-of-power system collapsed.[27]

While it was not seen during the Cold War, neorealists point out that small states might also be a determining factor in an actual balance-of-power situation if enough side with one great power over the other.[28] This might become relevant in the Asia-Pacific if China challenges the United States' hegemony. Almost all small states in the region currently side with the United States, although this could change if these small states attempt to play the two powers off against each other in exchange for military, economic, or other benefits. This is already occurring in the South Pacific, where microstates are using increased US and Chinese competition for influence in order to access development assistance and other benefits. After there was a military coup in Fiji in 2006, the United States and its partners introduced sanctions and attempted to isolate Fiji from the region and the international community. Fiji responded by adopting a "Look North" policy, whereby it sought closer relations with China and other Asian states. Fiji calculated that China's interest in gaining influence in the region would mean that it would respond favorably to Fiji's effort, and that in turn the United States and its partners would be forced to soften their approach. This tactic worked. Fiji accessed significant development and military assistance from China, as well as support in its international **diplomacy**, with China sponsoring Fiji to assume the chairmanship of the Group of 77 + China in 2013. To attempt to regain influence, the United States had to remove sanctions and build a new diplomatic mission in Fiji, the largest in the region. However, the importance of the South Pacific to China and the United States' regional interests are minimal, and any competition in the region is small scale. In the broader Asia-Pacific, we should expect less of this sort of behavior and influence from small states than from the middle powers, as discussed in chapter 4.

Liberalism is also useful for studying small states because, in contrast to realism, it argues that all states are different, so ideational factors (e.g., **norms**, identity, and ideas) are as important as material ones. Therefore, it is important to consider the particular characteristics of small states when analyzing their behavior, as they might not be driven by the same military and economic power calculus as great and middle powers, but instead by domestic considerations. For example, as described above, South Pacific microstates are especially vulnerable

Table 5.3 Recent Natural Disasters in the South Pacific

State	Date	Type	Fatalities
Solomon Islands	April 2007	Earthquake	52
Papua New Guinea	November 2007	Storm	172
Samoa	September 2009	Tsunami	143
American Samoa	September 2009	Earthquake	34
Papua New Guinea	January 2012	Landslide	60
Solomon Islands	April 2014	Flood	47

to natural disasters and to climatic and ecological threats, which shape their behavior on the world stage (table 5.3), particularly their pursuit of international action on climate change, as is discussed below.[29]

Given that small states have a limited individual capacity, **neoliberals** predict that small states will favor international and regional institutions to enhance their influence, because these institutions usually treat states equally. Membership in international institutions may allow small states to act "collectively to help shape developing international attitudes, dogmas, and codes of proper behavior."[30]

Regional institutions also allow small states to work together to pool their resources and operate as a united group when negotiating with larger states, which may enhance their bargaining power and influence. As is discussed in chapter 11, in the Asia-Pacific, the small Southeast Asian states have favored regional integration, exemplified by the creation of the **Association of Southeast Asian Nations (ASEAN)**, as well as its myriad associated organizations. Small states have worked together through ASEAN to enhance their influence when dealing with the great powers, the United States and China, on matters such as territorial disputes and trade agreements. During the Cold War these small states successfully worked together to prevent the region from becoming a great power playground. The South Pacific microstates have also favored regional institutions, particularly the **Pacific Islands Forum,** the main political institution in the region, and the **Pacific Community**, the most important developmental institution. Although Australia and New Zealand are members of the Pacific Islands Forum and the Pacific Community (along with France and the United States), South Pacific microstates are increasingly creating new regional and subregional institutions that exclude larger states, such as the **Melanesian Spearhead Group**, to enhance their influence when negotiating with them. While institutions can empower small states, neoliberals acknowledge that institutions can also exert significant influence over their foreign policies, as small states may be restrained by the interests of their allies' intentions or strategies within these institutions or by the expectations of the institutions themselves.[31] This might explain why the small states and microstates in the region are increasingly creating regional and subregional institutions that exclude larger states.

Small states are also likely to favor international law set by the United Nations, as it attempts to guarantee the sovereignty of states and therefore to protect small states from armed attack by larger states. The importance of international law is evident in the Asia-Pacific, where the small Southeast Asian states and the South Pacific microstates strongly advocate for the observance of international law. One of the most important international laws for the region's small states and microstates is the **UN Convention on the Law of the Sea (UNCLOS)**. UNCLOS sets the limits of each state's territorial waters and grants states sovereign rights for 12 nautical miles off their coastline. Within this limit, states are—in principle—free to enforce any law, regulate any use, and exploit any resource. UNCLOS also creates the concept of the **Exclusive Economic Zone (EEZ)**, within which coastal states have the right to exploit, develop, manage, and conserve all resources found in the waters, on the ocean floor, and in the subsoil of an area extending 200 miles from their shores.

UNCLOS is particularly important for small Southeast Asian states. As discussed in chapter 7, there are increasing disputes over the ownership of maritime territories in the region, especially in the South China Sea. UNCLOS provides a legal framework that small states can use to attempt to resolve these disputes. UNCLOS is also very important to South Pacific microstates. Given that South Pacific microstates consist of a multitude of islands and archipelagoes, UNCLOS grants them extensive EEZs. Indeed, many have EEZs that cover over 1 million square kilometers. As most South Pacific microstates have few natural resources and little industry, their major source of revenue is the fish that are caught in their EEZs. However, these microstates often struggle to patrol their EEZs, which leaves their fish stocks susceptible to exploitation by distant-water fishing nations. If these microstates were able to protect their EEZs and collect the full value of the fish caught in them, this could dramatically aid their development. In a related example of a neoliberal prediction regarding microstates' behavior, the South Pacific microstates have created a regional institution, the Pacific Islands Forum Fisheries Agency, which helps them to sustainably manage the fishery resources that fall within their EEZs and to work together when negotiating with larger states that want to fish in their territories.

A nontraditional security studies theory such as **constructivism** is also useful when analyzing small states, as it shows that, though material resources and institutions affect their security, ideational factors can also be influential. While small states do not have the military or economic power to exert significant influence over security issues in the Asia-Pacific, constructivism encourages us to analyze how they might be able to use norms, identities, and ideas to act as "norm entrepreneurs" in order to influence regional and international politics.[32] That is, small states might be able to mobilize support for particular standards and persuade other states (including great and middle powers) to adopt and conform to new norms.[33] Indeed, if they use their ideational power cleverly, small states "can provide a moral balance of power in the international system."[34]

In this regard, climate change poses a serious challenge to Asia-Pacific micro-states, particularly the island states of the South Pacific, as well as those small states that are prone to flooding, such as Bangladesh and Vietnam. Rising sea levels pose a threat to these island states, especially because many consist of coral reefs, atolls, and archipelagoes only a few meters above sea level. This may lead to food and water scarcity, as freshwater sources become contaminated with seawater and farming land is flooded. There is also evidence that natural disasters may increase with climate change. In 2008 Emanuel Mori, the president of the Federated States of Micronesia, argued before the UN General Assembly that "climate change also impacts our human rights. It impacts international peace and our own security, territorial integrity and our very existence, as inhabitants of very small and vulnerable island nations."

Because climate change poses such a significant threat to the security of South Pacific microstates, they have taken a leading role in attempts to achieve international action to address the issue. The degree of influence that these states have had at the international level has far outweighed their size and illustrates how small states and microstates can act as norm entrepreneurs. For example, in 2007 the UN Security Council (UNSC) debated the impact of climate change on peace and security. The representative of the Pacific Islands Forum, Papua New Guinea, persuasively argued that the impact of climate change on microstates was no less threatening than guns and bombs were to large states. In 2009 the UN General Assembly adopted a resolution—proposed by the South Pacific microstates—titled "Climate Change and Its Possible Security Implications" that called on the UNSC and relevant UN agencies to investigate the issue. At the 2009 Copenhagen UN Climate Change Conference, the South Pacific microstates worked within the Alliance of Small Island States to lead calls for greenhouse gas emission reduction targets to be agreed on. Although the Copenhagen talks did not result in significant improvements to the international response to climate change, the leading role taken by South Pacific microstates illustrates how small states can use ideational factors to enhance their influence.

Studying small states also invites us to draw on **critical security studies** to challenge the focus of traditional security studies theories on states as the primary referent of security. As described above, many small states (especially microstates) experience vulnerabilities that undermine their power and can contribute to their "weakness." To understand what a so-called weak state is, one first needs to understand what a "strong" or "working" state is. According to international law, a state has four characteristics: a defined territory, a permanent population, an effective government, and the capacity to enter into formal relations with other states.[35] There is no particular test of whether a government is "effective." The dominant approach owes its origins to the sociologist Max Weber, who defined an effective government as one "that (successfully) claims the *monopoly of the legitimate use of physical force* within a given territory."[36] This definition has since expanded to include the government's capacity to deliver public goods and services, which include security, law and order, and social goods such as infrastructure and health care.[37] However, an effective government also appears to require legitimate authority in order to motivate its citizenry to act as an organized, effective entity. Today, most Western states assume that a government is only legitimate if it is a liberal democracy.

Therefore, the answer to what a "weak" or "failing" state is has two dimensions. First, the legal dimension directs us to look at the extent to which the state is having difficulty satisfying the four characteristics identified above. Second, from a Western perspective, the political dimension directs us to look at the government of the state, and ask whether it is effective and legitimate. How the state is performing in each dimension will determine whether it is merely weak or is failing—or, indeed, has even failed. The classic definition is that states fail when "they can no longer perform the functions required for them to pass as states."[38] Therefore, critical security studies invites us to recognize that, in some small states that are weak, the security of the population might be more threatened by their state, and its failure to protect them, than by other states.

In this regard, after the end of the Cold War, the threat of nuclear war between superpowers was replaced by the threat of what the journalist Robert Kaplan famously described as the "coming anarchy" if "weak" (usually small) states "failed."[39] This saw the international community shift its security focus from "old" interstate wars to "new" intrastate ones.[40] Many of these wars took place in the Asia-Pacific's recently independent small states, such as Cambodia and Timor-Leste, which experienced weakness, and often failure, after independence. The threat posed by new wars in weak or failing states took on new urgency after the terrorist attacks of September 11, 2001, when **failed states** were identified as a major threat to international order and security, as more attention was paid to the fact that they can be vulnerable to humanitarian emergencies, and also potential breeding grounds or havens for terrorism and transnational crime.[41] The most notable example of this phenomenon was Afghanistan, which before the 9/11 attacks had been a weak state controlled by the Taliban, a fundamentalist political movement that supported the terrorists who conducted the attacks. As a result, state weakness and state failure were "securitized," a process identified by the **Copenhagen School**. For example, the 2002 US *National Security Strategy* stated that "America is now threatened less by conquering states than we are by failing ones."[42]

A number of Asia-Pacific states can be considered weak, although arguably none has failed. For example, though Cambodia has gone through a number of changes since the 1980s—including transitioning from war to peace, from economic isolation to integration, and from one-party rule to multiparty democracy—its government remains weak, for it is hobbled by corruption and a poor relationship with its society. Similar challenges are evident in Laos, Burma/Myanmar, Nepal, and Sri Lanka.

Weak or failing small states may require assistance from larger states to strengthen their capacity. In these circumstances, larger states may engage in "**state building**," during which they attempt to help a small state achieve control over its territory, gain the loyalty of its population, and build durable, centralized institutions that hold a monopoly over violence. In some circumstances, this assistance might be requested by the small state; but in others, larger states might decide to intervene when the small state is unable, or unwilling, to make such a request. If a small state is enduring a security crisis and/or a humanitarian emergency, the narrow school of the **human security** approach described in the introduction and chapter 12 advocates measures to manage these threats, including **intervention**, that is, the threat or use of force by a state, group of states, or international organization against another state. The broad school of the human security approach adds to this by also advocating measures aimed at promoting development. Under Chapter VII of the Charter of the United Nations, the UNSC can authorize intervention to "maintain or restore international peace and security" (Article 42). The UNSC has authorized a number of interventions and state-building operations in the Asia-Pacific region, most notably the UN Transitional Authority in Cambodia (1992–93), the UN Transitional Administration in East Timor (and related missions, from 1999 to 2013), and the UN Assistance Mission in Afghanistan (2002–ongoing).

A number of interventions have also been undertaken in the Asia-Pacific on a bilateral or regional basis. For example, in 2003 the Solomon Islands requested, and the Pacific Islands Forum authorized, the Regional Assistance Mission to Solomon Islands (RAMSI). Authorization for RAMSI was sought from the Pacific Islands Forum because the Solomon Islands gives diplomatic recognition to Taiwan, which draws ire from China. Thus there was concern that China would veto the intervention if it was put to vote in the UNSC. This mission responded to ongoing conflict between the populations of the two largest islands in the Solomon Islands, Malaita and Guadalcanal, which had contributed to the weakness of the state and a deteriorating security situation. The initial deployment of RAMSI consisted of more than 2,000 personnel from Australia, New Zealand, Papua New Guinea, Tonga, and Fiji. Later on, Samoa, the Cook Islands, Vanuatu, Kiribati, Nauru, and Tuvalu also joined RAMSI. Rather than merely being an intervention to restore law and order and build the state, RAMSI adopted the broad school of the human security approach and was a long-term project aimed at promoting economic development, at an estimated cost of $2.4 billion during its first ten years.

Conclusion

This chapter has illustrated how studying small states offers a different way to analyze security in the Asia-Pacific, for it can challenge the assumptions made by the traditional theories. Traditional security theories such as neorealism and neoliberalism focus on the state-centric security challenges facing small states.

By contrast nontraditional theories such as constructivism, critical security studies, and human security reveal that, for many small states, the key security challenges are internal or nontraditional challenges such as climate change and state weakness. This suggests that you should not take for granted that the traditional security theories will always generate the most plausible predictions about which security challenges the region faces and how states will respond to them.

Therefore, this chapter has demonstrated that small states matter for security in the Asia-Pacific. When small states are strong, they are able to influence regional and international norms, such as those on climate change. When small states are weak, they can challenge the security of a larger state by offering havens to terrorists and transnational terrorists, often necessitating expensive interventions and state-building measures in response.

Key Points

- Although there are varying definitions of "small states," this chapter proposes that a small state is one that *has fewer economic, military, and societal resources than great and middle powers and that can consequently find itself vulnerable to interference by great and middle powers and difficult to act independently.*
- Neorealism predicts that small states will either bandwagon with great powers (illustrated by the San Francisco alliance system in the Asia-Pacific) or use balancing between two or more great powers to enhance their own security.
- Neoliberalism predicts that small states will favor international and regional institutions, as exemplified by ASEAN in Southeast Asia and the Pacific Islands Forum in the South Pacific, and international law, as illustrated by the priority placed on UNCLOS by Asia-Pacific small states.
- Constructivism highlights that small states can use ideational factors to act as norm entrepreneurs in order to enhance their international influence, as illustrated by the activities of the South Pacific microstates with respect to combating climate change.
- Critical security studies and the human security approach highlight how traditional state-centric security challenges may not be a priority for small states. Instead, internal challenges, including state weaknesses, may warrant international intervention to restore security.

Questions

1. What is the most persuasive definition of a "small state"?

2. How do you distinguish between "small states" and "microstates"?

3. Does neorealism provide a convincing account of how small states might seek to secure themselves?

4. Why do neoliberals argue that small states will favor international institutions and law?

5. How are nontraditional security studies theories useful for analyzing the security of small states?

Guide to Further Reading

Commonwealth Secretariat and World Bank. *Small States: Meeting Challenges in the Global Economy:* Report of the Commonwealth Secretariat–World Bank Joint Task Force on Small States. London: Commonwealth Secretariat, 2000.
This report provides a comprehensive review of the challenges and vulnerabilities experienced by small states.

Ingebritsen, Christine, Iver B. Neumann, Sieglinde Gstohl, and Jessica Beyer, eds. *Small States in International Relations.* Seattle: University of Washington Press, 2006.
This book represents a concerted attempt to theorize the security and foreign policy challenges by small states. It also consolidates key literature on this issue.

Kassimeris, Christos. "The Foreign Policy of Small Powers." *International Politics* 46, no. 1 (2009): 84–101. http://dx.doi.org/10.1057/ip.2008.34.
This article is a recent attempt to theorize the foreign policy of small states and contains a helpful review of the relevant literature.

Keohane, Robert O. "Lilliputians Dilemmas: Small States in International Politics." *International Organization* 23, no. 2 (1969): 291–310. http://dx.doi.org/10.1017/S00208 1830003160X.
This article represents one of the most prominent attempts to consider the foreign policy and security concerns of small states. It uses a liberal perspective to conduct this analysis.

Smith, Nicola, Michelle Pace, and Donna Lee. "Size Matters: Small States and International Studies." *International Studies Perspectives* 6, no. 3 (2005): ii–iii. http://dx.doi .org/10.1111/j.1528-3577.2005.215_1.x.
This short article outlines some of the key reasons why small states are worthy of consideration in the study of international relations.

Notes

1. Iver Neumann and Sieglinde Gstohl, "Lilliputians in Gulliver's World?" in *Small States in International Relations* ed. Christine Ingebritsen, Iver B. Neumann, Sieglinde Gstohl, and Jessica Beyer (Seattle: University of Washington Press, 2006), 3.

2. Ibid., 6.

3. Colin Clarke and Tony Payne, eds., *Politics, Security and Development in Small States* (London: Allen & Unwin, 1987); Commonwealth Secretariat and World Bank, *Small States: Meeting Challenges in the Global Economy: Report of the Commonwealth Secretariat-World Bank Joint Task Force on Small States* (London: Commonwealth Secretariat, 2000).

4. Simon Kuznets, "Economic Growth of Small Nations," in *Economic Consequences of the Size of Nations,* ed. E. A. G. Robinson (London: Macmillan, 1960).

5. Clarke and Payne, *Politics, Security and Development.*

6. Christos Kassimeris, "The Foreign Policy of Small States," *International Politics* 46 (2009): 84–101.

7. Philippe L. Hein, "The Study of Micro-States," in *States, Microstates, and Islands*, ed. Edward C. Dommen and Philippe L. Hein (London: Croom Helm, 1985).

8. Harvey W. Armstrong, R. J. De Kervenoael, X. Li, and R. Read, "A Comparison of the Economic Performance of Different Microstates and between Microstates and Larger Countries," *World Development* 26 (1998): 639–56.

9. Paul Collier and David Dollar, "Aid, Risk and the Special Concerns of Small States," paper presented at World Bank–Commonwealth Secretariat Conference on Small States, St. Lucia, February 17–19, 1999.

10. Commonwealth Secretariat and World Bank, *Small States.*

11. Michael Handel, *Weak States in the International System* (London: Frank Cass, 1981).

12. Armstrong et al., "Comparison."

13. Peter R. Baehr, "Small States: A Tool for Analysis?" *World Politics* 27 (1975): 456–66; Miriam Fendius Elman, "The Foreign Policies of Small States: Challenging Neorealism in Its Own Backyard," *British Journal of Political Science* 25 (1995): 171–217; Handel, *Weak States.*

14. Robert O. Keohane, "Lilliputians' Dilemmas: Small States in International Politics," *International Organization* 23 (1969): 291–310.

15. Baehr, "Small States"; Handel, *Weak States*; Baldur Thorhallson, *The Role of Small States in the EU* (Aldershot, UK: Ashgate, 2000).

16. Keohane, "Lilliputians' Dilemmas," 296.

17. Commonwealth Advisory Group, *A Future for Small States: Overcoming Vulnerability* (London: Commonwealth Secretariat, 1997), 13; United Nations, *Development of a Vulnerability Index for Small Island States*, Report of the Secretary-General, UN Document A/53/65–E/1998/5 (New York: United Nations, 1998).

18. Commonwealth Advisory Group, *Future for Small States*, 13.

19. Ibid.

20. Lino Briguglio, Gordon Cordina, Nadia Farrugia, and Stephanie Vella, *Economic Vulnerability and Resilience: Concepts and Measurements*, Research Paper 2008/55 (Helsinki: World Institute for Development Economics Research of United Nations University, 2008).

21. Robert Jervis, "Cooperation under the Security Dilemma," *World Politics* 30 (1978): 167–214.

22. Stephen M. Walt, *The Origins of Alliances* (Ithaca, NY: Cornell University Press, 1987).

23. Benjamin Miller and Korina Kagan, "The Great Powers and Regional Conflicts: Eastern Europe and the Balkans from the Post-Napoleonic Era to the Post–Cold War Era," *International Studies Quarterly* 41 (1997): 51–85.

24. Kassimeris, "Foreign Policy," 94.

25. Handel, *Weak States.*

26. Efraim Inbar and Gabriel Sheffer, "Introduction," in *The National Security of Small States in a Changing World*, ed. Efraim Inbar and Gabriel Sheffer (London: Frank Cass., 1997).

27. Jeanne A. K. Hey, "Introducing Small State Foreign Policy," in *Small States in World Politics: Explaining Foreign Policy Behaviour*, ed. Jeanne A. K. Hey (Boulder, CO: Lynne Rienner, 2003).

28. Nicola Smith, Michelle Pace, and Donna Lee. "Size Matters: Small States and International Studies," *International Studies Perspectives* 6 (2005): ii–iii.

29. Commonwealth Secretariat and World Bank, *Small States*.

30. Keohane, "Lilliputians' Dilemmas," 297.

31. Kassimeris, "Foreign Policy."

32. Christine Ingebritsen, "Norm Entrepreneurs: Scandinavia's Role in World Politics," *Cooperation and Change* 37 (2002): 11–23.

33. Martha Finnemore and Kathryn Sikkink, "International Norm Dynamics and Political Change," *International Organization* 52 (1998): 887–917.

34. Christine Ingebritsen, "Learning from Lilliput," in *Small States*, ed. Ingebritsen et al.

35. Montevideo Convention on the Rights and Duties of States, 1933, Article 1.

36. Max Weber, "Science as a Vocation," in *From Max Weber: Essays in Sociology*, ed. Hans H. Gerth and Charles Wright Mills (New York: Oxford University Press, 1958), 78.

37. Robert I. Rotberg, "Failed States, Collapsed States, Weak States: Causes and Indicators," in *State Failure and State Weakness in a Time of Terror*, ed. Robert I. Rotberg (Washington, DC: Brookings Institution Press, 2003).

38. William I. Zartman, ed., *Collapsed States: The Disintegration and Restoration of Legitimate Authority* (London: Lynne Rienner, 1995), 5.

39. Robert Kaplan, "The Coming Anarchy," *The Atlantic*, February 1994, http://www.theatlantic.com/magazine/archive/1994/02/the-coming-anarchy/304670/.

40. Mary Kaldor, *New and Old Wars* (Stanford, CA: Stanford University Press, 2007).

41. Jeremy Weinstein, John Edward Porter, and Stuart E. Eizenstat, *On the Brink: Weak States and US National Security* (Washington, DC: Commission on Weak States and US National Security, 2004), 1.

42. White House, *The National Security Strategy of the United States of America* (Washington, DC: White House, 2002).

Part II

Current and Emerging Security Challenges

6 Military Modernization and Arms-Racing in the Asia-Pacific

Tim Huxley and Brendan Taylor

Reader's Guide

On the back of four decades' worth of largely uninterrupted economic growth, many if not most states in the Asia-Pacific have been increasing their defense spending and military equipment procurement. Today, states in the region lead the world collectively by accounting for approximately half of all global arms imports, far ahead of any other region. Yet there has been much debate over whether this phenomenon principally reflects a traditional military modernization processes or whether it constitutes something more worrying and potentially destabilizing in the form of an emerging Asia-Pacific "arms race."

Exploring this debate, this chapter is divided into three sections. The first examines three major controversies surrounding the concept of an "arms race," centering on the definition of this concept, as well as the *causes* and *consequences* of arms races. The second section provides the reader with some general background on the military programs being undertaken by several governments and their armed forces (those of the United States, China, Japan, India, Vietnam, and Indonesia) in the Asia-Pacific. The third section then draws together the first two sections, providing an overview of the debate over whether there is an arms race unfolding in the Asia-Pacific, and highlighting the reasons why scholars and analysts have produced such different answers in response to this question. As well as acknowledging that there remains much room for debate about whether contemporary regional military developments should be classed as an arms race, the chapter concludes by exploring the ramifications of these developments for those interested in Asia-Pacific security, whether as scholars or as policymakers.

What Is an Arms Race?

Notwithstanding its prevalence in the lexicon of security and strategic studies—and, indeed, among policymakers and in broader public discourse—the notion of the "arms race" is highly contested and potentially confusing. Indeed, some prominent scholars have even advocated abandoning the term altogether.[1] Three important academic debates surround the arms race concept. The first of these debates is *definitional*. A major point of contention here concerns where semantic and analytical boundaries should be drawn. How can arms-racing behavior, for instance, be distinguished from the military modernization processes in which many states engage? The second debate is about the *causes* of arms races. Here a key point of difference is between those scholars who argue that arms races, on balance, are caused more by external factors and those at the other end of the analytical spectrum who contend that arms races are driven largely by domestic political dynamics. A third debate concerns the *consequences* of arms races. This debate can again be characterized in dichotomous terms, between those who argue that arms races are inherently destabilizing and those who contend that they reflect rational international behavior that can actually prevent conflict rather than fueling it.

Interstate dynamics that have been referred to as arms races have existed for a long time. The most famous historical case is that involving Britain and Germany in the lead-up to World War I. At the time, Britain "ruled the waves" as the world's leading naval power. This was a position that Germany sought to challenge by undertaking a major buildup of its relatively small navy into a major "blue water" fleet. Britain responded to this development in part by building a powerful new class of warship known as the "dreadnought," a design that the Germans subsequently copied. During the decade leading up to World War I, the number of dreadnoughts built by Britain was influenced significantly by the numbers built by Germany, and vice versa. This "arms race" dynamic has subsequently been regarded by many historians as a leading cause of World War I.[2] However, it was not until the onset of the Cold War and the subsequent birth of strategic studies as a field of academic study during the 1950s that scholars began systematically to study interstate arms racing dynamics. A large body of work emerged during this period, stimulated particularly by the prospect of an arms race—involving nuclear weapons and long-range missiles between the two superpowers, the United States and the USSR.

In one of the earliest and most influential contributions to the literature during this period, the Harvard University professor Samuel Huntington defined arms races as "a progressive, competitive peacetime increase in armaments by two states or **coalitions** of states resulting from conflicting purposes or mutual fears."[3] According to Huntington's definition, the notion of "reciprocal interaction" was central to arms-racing behavior. By this, he meant that two states (or coalitions of states) needed to disagree over the "proper" balance of military power between them, and to be self-consciously increasing their arsenals—qualitatively or quantitatively, or both—specifically in the context of that disagreement. The leading strategist Colin Gray added later that, in order for arms-racing to occur, increases in weapons acquisitions also needed to occur rapidly. This was one important way of differentiating an arms race from otherwise routine military modernization.[4] To further emphasize this distinction, Huntington argued against the possibility of what he termed a "general arms race," where multiple states were arming simultaneously. In Huntington's view, such a dynamic was not the result of reciprocal interaction. At the very most, he considered it the sum of a series of two-state antagonisms.[5]

Despite the definitional efforts of Huntington, Gray, and others, use of the "arms race" terminology has only broadened with the passage of time. This has led to a degree of blurring and confusion between arms-racing behavior and the related, though still quite distinct, processes that many states undertake in attempts to maintain or improve their military capabilities, particularly in good economic times. Some scholars have sought to deal with this confusion by defining arms-racing behavior quite narrowly and, in the process, by articulating what *does* and what *does not* constitute an arms race. Barry Buzan and Eric Herring, for instance, developed a four-part analytical framework of what they term the "arms dynamic," of which "arms-racing" is one component. In their view, arms-racing is only present in those rare situations "when actors are going flat out or almost flat out in major investments in military capability."[6] They juxtapose this with what they term the "maintenance of the military status quo" to describe routine military modernization processes. In between these two extremes, Buzan and Herring include "arms competition/buildup," which refers to situations where adversaries are constantly seeking to improve their position relative to one another, but have no real confidence in their capacity to achieve a clear advantage. They also include "arms build down" as an analytical category used to capture those situations in which states dismantle particular weapons systems and replace these with smaller, less capable, and arguably less destabilizing platforms. Buzan and Herring's "arms dynamic" is itself quite flexible, leaving open as it does the possibility that a state's behavior and, indeed, interstate behavior can shift over time between these various places on an analytical spectrum.

Debates in the rather large literature on arms races have by no means been confined to such definitional disputes. Another main point of contention has centered on the factors that cause arms races. A prevalent view in the literature is that arms races are caused and sustained by predominantly *external* factors. Consistent with Huntington's classic definition of an arms race, this view assumes that states respond and react—largely out of fear and insecurity—to the arms acquisitions of an adversary. This reaction, in turn, prompts the adversary to respond in kind—again out of its own sense of fear and insecurity—starting a potentially endless spiral that might culminate in open conflict. A second variant of the "*external* causes" school of thinking suggests that arms races are started by states that are not insecure, but are simply adventurist and arm themselves with a view to wresting additional territorial resources away from other states. While the Anglo-German naval race may, and indeed has been, understood in terms of the action/reaction model of arms racing, an alternative explanation is that it was fundamentally driven by Germany's desire to acquire additional colonial territories. Yet despite the obvious differences between these two schools of thought, both assume that states are rational and that they will respond in largely predictable ways to external stimuli.[7]

Juxtaposed against the "external causes" school is an opposing body of work, which contends that arms races are sparked and sustained by domestic political factors in the states concerned. Such factors include the role played by electoral politics, wherein politicians opt to fund weapons projects that are likely to deliver economic gains to constituencies that will subsequently provide electoral support for them or their political party. Another argument is that arms races may be caused by bureaucratic politics, meaning that they are the product of jostling and competition between various parts of government, each seeking to advance its organizational interests. Because of the lengthy and technical nature of weapons acquisition, armed forces are often influential actors in such processes, and on occasion different military branches may compete with one another to press for the acquisition of particular systems or platforms relevant to their service. Regardless of the domestic factors at play, a key unifying strand in the "*internal* causes" school is that arms races are typically not the result of states reacting rationally to external stimuli. Instead, this school suggests that arms races are the product of "strategic autism," wherein state behavior is driven more by internal stimuli than by external incentives or threats.[8]

A third relevant academic debate concerns the *consequences* of arms races. This one can also be categorized in terms of two broad schools of thinking. The first suggests that arms races are inherently destabilizing phenomena that increase the probability of war. Advocates of this school have been prominent for at least the last century. In the aftermath of World War I, for instance, scholars seeking to make sense of that calamity identified the military buildup that had occurred in the decades leading up to the outbreak of war in 1914 as a major cause of the conflict due to the spiral of insecurity that the pre-1914 arms race engendered. Other scholars contest this interpretation, however, arguing that arms-racing is a quite rational form of state behavior that may also have a stabilizing effect. According to this line of thinking, arms-racing behavior can maintain a state of equilibrium or "balance of power" in the international system, particularly among the great powers. Others have argued that arms-racing behavior may even be stabilizing, because it encourages states to cooperate. The influential strategic commentator Bruno Tertrais, for instance, suggests that "to the extent that arms races create arms control, and arms control fosters confidence building, then arms races ultimately can have a positive effect."[9] Between these two ends of the analytical spectrum, some scholars have sought to place a foot in each camp by demonstrating that the *consequences* of arms races are context-specific. Huntington's seminal article from the late 1950s remains influential here. In that piece, he argues that the longer an arms race runs, the less likely it is to prove destabilizing. This is because, in Huntington's terms, "the sustained regularity of the increases in itself becomes an accepted and anticipated stabilizing factor in the relations between the two countries."[10]

Beyond the three debates outlined here, a further factor complicating the analysis of arms races is that such behavior may be related to the larger conceptual and theoretical approaches to security—such as the "security dilemma"—that are outlined in the opening chapter of this book. Moreover, the causes and consequences of arms-racing behavior may look quite different, depending upon the conceptual lens through which they are viewed. Realist scholars, for instance, might argue that arms-racing is a completely rational and possibly even a stabilizing fact of life in an anarchic international system in which states pursue their national interest by seeking to maximize their military power. Liberal institutionalists, by contrast, will view arms-racing behavior as symptomatic of a lack of international trust that needs to be mitigated through the formation of multilateral organizations and other associated processes, such as arms control agreements. Constructivist scholars may take yet another view, arguing that the significance of arms races is ultimately subjective. To invoke Alexander Wendt's well-worn phrase, "A gun in the hands of a friend is a different thing from one in the hands of an enemy."[11]

Military Programs in the Contemporary Asia-Pacific

The first section of this chapter has differentiated arms-racing behavior from the military modernization programs in which states engage. It is important, however, to recognize that arms races are still a component of these larger military modernization processes, as Buzan and Herring demonstrate in their work. Understanding these larger modernization processes is thus essential to identifying whether arms-racing behavior is occurring, and this section provides concise overviews of the military development programs currently being undertaken by the region's four major powers—the United States, China, Japan, and India—as well as those of two of rising Southeast Asian middle powers—Vietnam and Indonesia (figure 6.1).

The United States

Although the United States still spends more than any other country on defense, and indeed accounts for roughly half the world's total military expenditure, its government has faced serious funding constraints since the 2008–9 global financial crisis. US military spending increased rapidly during the decade following the September 11, 2001, terrorist attacks, but subsequently the defense budget was threatened by significant cuts. After peaking at $720 billion in 2011, US military

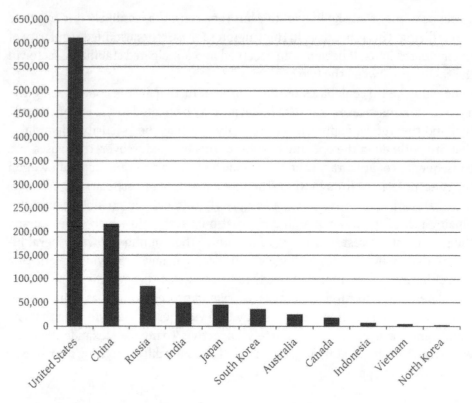

Figure 6.1 Military Spending of Selected States in Asia (in millions of dollars), 2014
Note: The data for China, Russia, and South Korea are estimates. Reliable data for North Korea and Vietnam are not available because military spending is not transparent, so SIPRI estimates are based on third-party and incomplete information.
Source: Stockholm International Peace Research Institute, "SIPRI Military Expenditure Database."

spending was expected to decline to approximately $580 billion in 2015.[12] An increasing proportion of this overall US defense effort is being directed toward the Asia-Pacific, as part of the United States' "rebalance" to the region since 2011. At the International Institute for Strategic Studies' Shangri-La Dialogue in June 2013, US secretary of defense Chuck Hagel announced that 60 percent of US naval assets will be based in the Pacific region by 2020, along with 60 percent of the US Air Force's overseas-based forces and "a similar percentage" of its space and cyber capabilities.[13] The United States is also strengthening its ground forces in the region, notably through the deployment of US Marines to Darwin in northern Australia. New concepts, **doctrines**, and plans are being developed to integrate new technologies and capabilities, including remotely piloted vehicles and directed-energy weapons, to "ensure freedom of action throughout the region well into the future."[14]

The Obama administration has consistently denied that rivalry with China has motivated the United States' "rebalance" to the Asia-Pacific. It has sought both to maintain equable relations with Beijing and to reassure Southeast Asian governments that the strong military and security element of the United States' new interest in their region does not indicate the beginnings of a new cold war that might ultimately force them to take sides. Nevertheless, it is clear that a reaction to China's growing power, confidence, and assertiveness in a part of the world that Washington assesses to be strategically important was a core motivation for the rebalance. It is also evident that the United States' "Air-Sea Battle" concept, which included potential offensive operations against targets in adversaries' homelands, as well as in space and cyberspace, was intended to help the US armed forces offset China's military buildup and its potential to complicate US military operations in the Asia-Pacific region. Consistent with this, some critics of Air-Sea Battle were concerned by its potential to stimulate enhanced Chinese efforts to develop "anti-access / area denial" (A2/AD) capabilities, and to place a premium in any future conflict on preemptive strikes, which could be followed by rapid escalation.[15]

China

China's efforts to improve its military capabilities have been wide-ranging and have reflected a variety of factors. These include internal drivers such as the country's continued rapid economic growth and the important relationship between the ruling Communist Party and the People's Liberation Army (PLA). External issues have also pushed the development—including the unresolved issue of Taiwan's political status; growing tensions over conflicting territorial claims in the East China Sea and South China Sea; and, according to Beijing's latest defense white paper in 2013, a requirement for armed forces "commensurate with China's international standing."[16] Chinese sources also express concern over the military dimensions of the United States' rebalance to the Asia-Pacific, which has been widely interpreted in Beijing as a thinly disguised effort to contain and balance China's growing power and strategic extroversion.

Because Beijing's official figures do not provide a full picture, arriving at an accurate estimate of China's defense spending is not straightforward. However, this spending has grown quickly since 2001, by double-digit percentages in most years, more or less in tandem with the rapid expansion of China's economy. China's official defense budget in 2014 was $129 billion. In early 2015 reports indicated that this budget was set to rise by a further 10 percent compared with the previous year, meaning that defense expenditures are continuing to outpace China's economic growth.[17] While this level of military spending means that China has the world's second-largest defense budget, Beijing still spends significantly less on defense than the United States.

Efforts to expand the PLA's capabilities have been comprehensive, but those relating to the People's Liberation Army Navy (PLAN) have been particularly notable. In 2012 the PLAN commissioned its first aircraft carrier, the *Liaoning*.[18] The PLAN expects the carrier to reach full operational capability after four to five years of trials, and has plans to develop a full carrier battle group. A second carrier may enter service by 2020.[19] At the same time, a new class of large amphibious warfare ships—capable of carrying transport, antisubmarine warfare, and attack helicopters as well as more than 1,000 marines—is also reportedly under development.[20] These capital ship projects highlight China's growing investment in maritime power projection capabilities alongside the widely recognized PLA emphasis on establishing an effective A2/AD capacity. These programs are intended to make it more difficult for the United States to intervene in regional crises.[21]

Another clear priority for China has been to develop its nuclear and space forces, including long-range missiles, in the form of the land-based DF-41s and the submarine-launched JL-2s, which could reach the continental United States. During 2013 China carried out its second missile-interceptor test, prompting speculation that it was developing its own missile defense system.[22] A US defense official has claimed that another Chinese missile launch was the first test of a new interceptor designed to destroy satellites in orbit.[23] The PLA has also been prioritizing the development of its cyberwarfare capabilities: in early 2013, the cybersecurity firm Mandiant claimed that it had identified a specific PLA unit as the source of extensive cyber espionage aimed at the United States and other countries.

Japan

China's assertion of its claims against the Japanese-administered Senkaku/Diaoyu Islands in the East China Sea, along with continuing concern in Tokyo over North Korea's nuclear and missile programs and aggressive behavior, have contributed significantly to Japan's more assertive posture on security matters, particularly since Shinzō Abe's election victory in 2012. Japan already possessed armed forces that were in some respects the best equipped of any Asian state (e.g., their submarines), despite the tradition of spending no more than 1 percent of gross domestic product on defense. In 2013 the defense budget amounted to $51 billion, after the first increase for eleven years.[24] Abe's government has moved—albeit with considerable care—to reinterpret the Constitution in order to allow Japan to engage in "collective self-defense" (in other words, joint operations with the United States and other partners).[25] Abe has also emphasized the need for a "dynamic defense force," which involves the continued reorientation of the Ground Self-Defense Force toward the southwest of Japan, as well as an emphasis on developing amphibious warfare capability.[26]

In 2014 Japan's defense budget included provision for extra surveillance capabilities in the southwestern islands, and accelerated training for a new amphibious warfare unit. Procurement projects for the 2014–19 period, outlined in the five-year Mid-Term Defense Plan released in December 2013, call for acquisition of Global Hawk unmanned air vehicles for long-range surveillance, F-35A Joint Strike Fighters, V-22 tilt-rotor transport aircraft, new destroyers, and additional submarines.

However, without a significant increase in defense spending, efforts to strengthen Japan's defense are unlikely to be realized. That said, starting in April 2014, the Abe administration's easing of Japan's self-imposed ban on arms exports may help strengthen the country's military capabilities in the longer term by allowing joint development of military equipment with the United States and other partners, and by securing economies of scale in production to the benefit of Japan's forces.[27] Abe's reelection in December 2014, as the result of a snap election portrayed as a litmus test of his reformist economic policies, is likely to go some way toward reinforcing Japan's military "normalization." Indeed, Abe began his new term in December 2014 by reiterating his intent to revise Japan's Constitution and the military constraints that it imposes.[28]

India

India's defense budget for 2015 will increase to approximately $40 billion, making it the world's eighth largest, after those of the United States, China, Saudi Arabia, Russia, the United Kingdom, France, and Japan. India has also for several years now been the world's largest importer of defense equipment.[29]

Continued emphasis on reequipping India's army has reflected enduring concern in New Delhi over the need to deter and potentially defeat land-based threats, from both China and Pakistan. Heavy armor units are being equipped with T-90S main battle tanks, earlier versions of which will be modernized.[30] Other priorities for army procurement include replacing its aging short-range, USSR-supplied air-defense missiles,[31] and acquiring new light helicopters.[32] Concern over China's growing capabilities has more directly influenced Indian thinking about air force modernization. However, progress toward acquiring new aircraft for the Indian Air Force's combat fleet has been painfully slow. The indigenously developed Tejas LCA (light combat aircraft) is now edging toward service entry, but—as has become normal with defense equipment developed and produced in India—lengthy delays and cost overruns have plagued the project, which is lagging fifteen years behind schedule. Longer-term Indian combat aircraft plans involve the Fifth-Generation Fighter Aircraft, which is intended to be a locally produced version of the Russian Sukhoi T-50 design, and the indigenous fifth-generation Advanced Medium-Combat Aircraft.[33] However, the disappointing track record of India's aircraft industry suggests that these projects may be slow to materialize.

It is India's naval program that reflects its nascent strategic competition with China most clearly. It has been a matter of pride for India to deploy an indigenously built aircraft carrier before China does so, and in August 2013 the hull of the 40,000-metric ton carrier *Vikrant* was launched. *Vikrant* is expected to enter service in 2018, and long-term plans call for a second, larger indigenous carrier, possibly nuclear-powered.[34] In the meantime, the Russian-built carrier *Vikramaditya* arrived at her home port of Karwar in southwestern India in January 2014.[35] India has also prioritized the development of its nuclear submarine force, with long-term plans calling for at least three nuclear-powered ballistic-missile boats (known as SSBNs) and six or more nuclear-powered attack boats (known as SSNs).[36] While there are serious shortcomings in India's conventional submarine force, which is operating at only two-thirds the strength envisaged in the navy's earlier long-term plans, six French-built Scorpene submarines are scheduled to enter service from 2016.[37]

Programs involving new destroyers, frigates, and offshore patrol ships are significantly expanding the Indian Navy's surface capabilities, while the first deliveries of United States–supplied P-8I antisubmarine warfare and maritime patrol aircraft, and the introduction into service of MiG-29K carrier-borne combat aircraft (which will equip the INS *Vikramaditya*), are strengthening naval air power.[38] Strategic competition with China, combined with the new Modi government's increased defense spending, will almost certainly ensure continuing ambitious naval procurement. Meanwhile, the increased opportunities for foreign investment in India's defense industry under the new administration may also help to boost India's military capabilities in the medium term and beyond.

Russia

Since 2008 Russia has been modernizing its armed forces, paid for through a significant boost in spending to $70 billion, part of a 25 percent increase in real terms.[39] This has involved significant new arms acquisitions, along with efforts to overcome sustained shortcomings in manpower, training, and serviceability of military equipment.[40] While the boosts in special forces and elite troops have troubled Europeans after the annexation of Crimea and invasion of Ukraine, the countries of Asia are most focused on Russia's Pacific Fleet.

The Russian Navy has also grown thanks to the modernization push of recent years. In 2013 it put into service a new *Borei*-class SSBN, and there are five more planned for introduction in coming years. Their primary focus will be to strengthen the country's strategic nuclear deterrent capability. There is also a significant push for expanded surface combatant production. This represents an "ambitious naval re-armament program over the next 20 years."[41]

However, the Pacific Fleet is not the primary focus for Russia or even its Navy. The bulk of these new vessels may be allocated to the Northern Fleet and European theater. Military analysts have substantial doubts as to when and in what state these capabilities will emerge. The end of the Cold War and USSR spending badly damaged the Russian Pacific fleet, and there are real discrepancies between the on-paper numbers of the fleet and the number operable for exercises and use on behalf of state interests.[42] In 2015 economic pressure from a falling currency and economic sanctions also put a hold on many of Russia's military modernization plans.[43] The role of the Russian fleet and armed forces in Asia is therefore likely to remain one of strategic deterrence, despite the grand plans.

Indonesia

In Indonesia, fast economic growth has allowed the government to direct increased resources toward defense spending, which amounted to $7.8 billion in 2013, up from $5.7 billion a decade previously.[44] Military capability improvements under the Defense Strategic Plan 2024 are shaped by the idea of establishing a "Minimum Essential Force" for the country's defense against external and domestic threats. Indonesia's emphasis has been strengthening naval and air capabilities. This reflects Jakarta's concern over rising tensions in the South China Sea and major power competition in the Asia-Pacific (in the context of a long-standing ambition to provide more effectively for the defense of the waters and airspace of the massive Indonesian archipelago), as well as a politically motivated desire to balance the army's traditional dominance of Indonesia's armed forces.

Indonesia's single most important defense procurement program involves submarines, with an eventual total of ten planned to be operational by 2024. Jakarta ordered an initial three German-designed type-209–1200 submarines from South Korea in 2011; the first and second boats are scheduled for delivery by 2017, and the third (which will probably be built in Indonesia) in 2019–20. Further submarines will not necessarily be of the same type, and reports indicate that Jakarta is considering buying *Kilo*-class boats from Russia.[45] New surface ships on order or entering service include two additional *Sigma*-class corvettes (to be partially built in Indonesia) ordered in 2012–13, and three British-built corvettes originally constructed for Brunei.

Meanwhile, Indonesia's air force took delivery in September 2013 of the final two of six Su-30 MK2 combat aircraft ordered from Russia in 2011, bringing its total inventory of Su-27/30 aircraft to sixteen. Indonesia began taking delivery of twenty-four refurbished ex–US Air Force F-16C/D fighters in mid-2014, with the order expected to be completed by the end of 2015. In September 2013 the army received its first two *Leopard* 2 tanks and two *Marder* infantry fighting vehicles from total orders of 103 and 50, respectively.[46] Visiting Jakarta in August 2013, Defense Secretary Hagel confirmed that the United States would sell eight AH-64E attack helicopters to Indonesia.[47]

Vietnam

While rapid economic growth has allowed Hanoi to increase its defense spending to $4.264 billion in 2014, its military budget remains relatively small compared with those of other Asia-Pacific countries.[48] Nevertheless, improving naval and air force capabilities are an important objective for Vietnam, one of the main contestants of China's claims in the South China Sea. In April 2013 Vietnamese navy personnel began training in Russia on the first of six Project 636E *Kilo*-class submarines ordered in 2009; the first two boats were delivered to Vietnam in 2014, with the remainder due by 2016.[49]

In August 2013 the Dutch company DSNS announced that it had reached an agreement with Vietnam on the supply of two *Sigma*-class corvettes (though it was widely reported that the contract was for four such ships).[50] Reports in February 2014 indicated that Hanoi had ordered an additional two *Gepard*-class (Project 11661K) frigates from Russia, boosting the eventual inventory of these ships to six.[51] Together, the new Dutch and Russian ships promise to boost Vietnam's surface combatant fleet significantly. Also in August 2013, Vietnam pursued its incremental air force modernization by ordering an additional twelve Su-30 MK2 combat aircraft from Russia.[52]

Is There an Asian Arms Race?

It is clear that many states in the Asia-Pacific region—including Australia, South Korea, and some other Southeast Asian actors, notably Singapore, whose defense programs are not examined here—are increasing their defense spending, procuring advanced military equipment, and making efforts to improve their military capabilities, particularly in the maritime domain (figure 6.2). But the implications of this for regional security are unclear. However, the snapshot of the military programs of some important Asia-Pacific powers presented here raises the question of whether such developments are consistent with arms-racing. This is a hotly contested subject among those interested in Asia-Pacific security, whether from an academic viewpoint or as practitioners.

One of the most important contributors to this debate in recent decades has been Desmond Ball of the Australian National University. The evolution of Ball's assessments and analysis over the last two decades highlights how developments in regional states' military programs in the last several years have reinforced the credibility of the idea that an arms race is indeed now under way. Ball's original work on the theme, published in 1993, examined what lay behind "alarmist rhetoric" in the press at that time about a supposed Asian arms race. Ball examined trends in regional states' military spending and equipment acquisitions, and he attempted to explain these developments by assessing the wide range of factors influencing them. He surveyed in detail not only the diverse influences affecting the acquisition of major weapons systems but also the major "themes" in regional states' military procurement programs: command, control, and communications systems; technical intelligence systems; multi-role fighter aircraft; maritime reconnaissance aircraft; modern surface combatants; antiship systems; electronic warfare capabilities; and rapid deployment forces.[53]

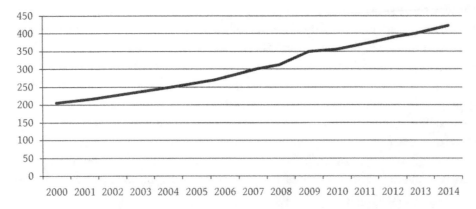

Figure 6.2 Total Military Spending in Asia (in billions of dollars), 2000–2014
Source: Stockholm International Peace Research Institute, "SIPRI Military Expenditure Database."
Asia includes Oceania and East, Central, and South Asia, but not the Americas.

In 1993 Ball concluded that there was no regional arms race. He argued that the two principal features characteristic of previous arms races—"a very rapid rate of acquisitions" and "some reciprocal dynamics"—were not obviously present in the region's military acquisition programs. Nevertheless, he highlighted two "disturbing" aspects that needed to be addressed if they were not to "overwhelm the more positive aspects of the emerging post–Cold War security architecture in the region." The first was the regional atmosphere of "uncertainty and . . . lack of trust" that was contributing to tensions and misunderstandings. The second element was the "offensive" character of some weapons systems that regional states were acquiring; Ball pointed to the strike capabilities associated with maritime attack aircraft, surface ships, and submarines, "all equipped with antiship missiles." These capabilities, along with new fighter aircraft, were "the most likely to generate counteracquisitions." Particularly where submarines and long-range antiship missiles were concerned, Ball pointed to worrying implications for crisis stability, and the danger of "inadvertent escalation."[54]

Returning to the regional arms dynamic topic in a paper in 2011, Ball presented a much more pessimistic outlook on regional military trends, focused on Northeast Asia. He argued that this subregion, which was "wracked with interstate tensions and disputes," was now "strategically the most worrisome . . . in the world." In his view, the chance of one or more of these "degenerating into large-scale conflict" with "horrendous consequences" was "palpable."[55] Ball surveyed the daunting range of disputes bedeviling Northeast Asia's international relations (including the existential tensions over Taiwan and North Korea, as well as acute territorial disputes involving almost every combination of states in the subregion), and assessed defense spending trends in Northeast Asia, pointing out that China, Japan, South Korea, North Korea, and Taiwan jointly accounted for more than 80 percent of total military expenditures in East Asia and Australasia. He particularly noted China's increasing defense expenditures. He highlighted the way these substantial defense budgets were being used in part to fund a "sustained and rapid buildup of defense capabilities." In contrast to the 1990s, Ball argued that there was now "substantial evidence of action/reaction dynamics" and specifically "an emerging complex arms race in Northeast Asia." Ball described in detail three main elements of the emerging subregional arms race in this subregion: naval capabilities (including the proliferation of major surface combatants, submarines, and antisubmarine warfare capability, and large amphibious ships); electronic warfare and electronic intelligence; and nuclear weapons and missile developments, including antimissile systems. He also analyzed in detail the competitive development of Northeast Asian states' information warfare (and more specifically cyberwarfare) capabilities.[56]

Ball argued that because the action/reaction phenomenon "generates its own momentum" and because "there are no arms control regimes whatsoever in Asia that might constrain or restrain" future military developments, an arms race is becoming increasingly prominent in Northeast Asia. Further, he pointed out that the region's future security dynamics were likely to be "much more complex" than those of the Cold War, noting for example that Chinese and North Korean ballistic missile developments have led Japan and South Korea to greatly enhance their airborne intelligence collection and early warning capabilities as well as their theater missile defenses, while US doctrine has evolved to allow the near-simultaneous employment of nuclear forces, precision conventional capabilities, and information warfare operations, and "the use of nuclear weapons in otherwise non-nuclear situations."[57] At the same time, the "distinctive categories, milestones and firebreaks which were carefully constructed during the Cold War to constrain escalatory processes and promote crisis stability" were missing. Ultimately, "in this environment, with many parties and many levels and directions of interactions, the possibilities for calamity are high."[58]

Robert Hartfiel and Brian Job, two scholars at the University of British Columbia in Canada, have also argued that there is an arms race—or what they more circumspectly term "competitive arms processes"—under way in Asia. Hartfiel and Job take as their starting point some of the analyses produced during the 1990s (which Ball contested at the time), characterizing this work as "prescient." They argue that the effects of the 1997–98 Asian financial crisis, which was seen to discredit the arguments of those who pointed to an emerging Asian arms race, were actually quite short-lived. In the period since the 9/11 terrorist attacks and possibly even earlier, they make the case that East Asian military modernization was again intensifying, and that the start of the so-called war on terror reinforced this trend by building public support for increased defense spending in some Asia-Pacific countries. A large part of the case they put forward for the existence of an Asian arms race, however, relates to the types of weapons platforms being acquired—such as fighter aircraft, surface combatants, submarines, and missiles—and their supposedly destabilizing character. A further destabilizing element, in their view, was the acquisition of these weapon systems by regional states that have historically been rivals. For these reasons, Hartfiel and Job concluded that "the troubling combination of volatile political conditions and destabilizing weapons increases the risks of both the accidental and deliberate outbreak of war."[59]

By contrast, others have adopted a more cautious view of regional military developments and whether these constitute an arms race. The naval historian Geoffrey Till, for instance, has also looked in detail at contemporary military trends in the Asia-Pacific, focusing on their maritime dimension. In his major 2012 study, Till examined the naval programs of China, India, Japan, and the United States with a view to discerning whether or not the Asia-Pacific was on the brink of a naval arms race. He argued that assessing whether there is an arms race requires looking beyond simply the acquisition of arms; it was important, he said, to assess the "type of arms procured and the doctrinal thought within a navy that lies behind them."[60] He found evidence in the contemporary Asia-Pacific of at least some of the classic characteristics of arms races; like Ball, he saw such characteristics as most apparent in Northeast Asia. However, Till argued that though substantial modernization was occurring, some of this was the result of internal rationales, such as the domestic political advantage accruing to politicians from military modernization, particularly where this produced economic benefits. Till concluded that the evidence for a naval arms race in the contemporary Asia-Pacific was therefore "mixed." While there was evidence of "elements of arms racing" involving the Chinese, Indian, Japanese, and US navies, and particularly between China and the United States, he argued that evidence against the existence of a regional naval arms race included the region's growing "sea-based economic inter-dependence" and the interest of all parties in naval cooperation against "common threats" such as natural disasters and piracy.[61]

Like Hartfiel and Job, in an article published in 2014 Christian Le Miere focused on the types of weapons system that Asia-Pacific countries are procuring. He arrived, however, at a more sanguine conclusion regarding whether or not an arms race was occurring in the region. Le Miere argued that though there was clear potential for an arms race, such dynamics are not occurring at present because the nature of the weaponry being acquired, coupled with the intentions underlying those acquisitions, are largely *defensive* in nature. In Le Miere's view, Beijing's development of A2/AD capabilities, for instance, was intended to limit the capacity of the US Navy to operate in the East China Sea and South China Sea; it should not be seen as evidence of a Chinese attempt to attain complete control of these waters. Similarly, some other Asian armed forces developing A2/AD capabilities—for example, Japan and Vietnam—were attempting, in his assessment, to limit China's freedom to maneuver in regional waters. The interdependent nature of these acquisitions—coupled with the role that China's development of A2/AD capabilities has played in inspiring the United States' rebalance to the region and associated concepts such as "Air-Sea Battle"—clearly points to the existence of action/reaction dynamics. Rather than seeing this as evidence of an Asian arms race, however, Le Miere instead characterized these trends as an Asian "security quandary," because of the uncertainty of many Asian states (including China) about whether their largely defensive capability development programs will be sufficient to achieve desired strategic objectives.[62]

Richard Bitzinger also questions whether an Asian arms race is occurring, but focuses his attention squarely upon the Southeast Asian subregion. According to Bitzinger, at least two factors suggest that there is no arms race in this part of the world. First and foremost, he argues that the requisite features of an arms race as discussed earlier in this chapter—such as mutual antagonism and action/reaction weapons acquisitions—are not in evidence in Southeast Asia. Second, he points to the fact that the number of major weapon systems being purchased in this sub-region is relatively small. Nevertheless, he concedes that something more than "mere modernization" is taking place in Southeast Asia. Rather than a full-blown arms race, he characterizes this dynamic as an "arms competition." And although he regards this as falling short of an arms race, he cautions against complacency, on the ground that this Southeast Asian arms competition still has the potential to contribute to a quite destabilizing subregional security dilemma.[63]

So, as with the long-standing debates concerning arms-racing in general terms, there is much disagreement over whether contemporary military modernization trends in the Asia-Pacific constitute an arms race. As shown here, answers to this question may differ depending upon which part of the region is being analyzed. Where Ball argues that the classic characteristics of an arms race are apparent in Northeast Asia, Bitzinger—like Ball himself—finds less evidence to support this conclusion in his study of Southeast Asian military acquisitions. Likewise, examination of particular *types* of military capability being developed and differing assessments regarding the *motivations* underpinning these efforts have also led to divergent conclusions. Hartfiel and Job, like Ball, argue that the weapon systems being acquired are predominantly offensive in nature, whereas Le Miere assesses the intentions underlying these capability development efforts as predominantly defensive. Similarly, harking back to wider conceptual debates about arms-racing, Till concludes that some of the military modernization occurring in Northeast Asia is being driven by internal rationales, thereby calling into question whether this is "arms racing" in the classic sense of the concept.

Conclusion

All the major powers, as well as many less powerful regional states in the Asia-Pacific region, are making efforts to improve their military capabilities. Whether this pattern of determined military modernization, which has resulted from diverse factors including economic success, constitutes an arms race is controversial. However, there is certainly evidence of action/reaction dynamics at work, and also of at least some of the key characteristics of past arms races.

One important caveat is that it may not be useful to assess the existence of an arms race in the early-twenty-first-century Asia-Pacific using templates developed from the historical examples of Anglo-German naval competition before World War I in Europe or the US-Soviet missile race during the Cold War of the mid–twentieth century. Arms races may have characteristics that are unique to their particular political and technological environments. In the contemporary Asia-Pacific, interstate tensions are complex and multipolar. At the same time, the complex military systems that the Asia-Pacific states are acquiring take considerably longer to develop and build than their twentieth-century predecessors. These systems are often much more lethal in terms of their destructive power and accuracy, implying that they be procured in relatively small numbers. Many key systems, such as those associated with "C4ISR" (i.e., command, control, communications, computing, intelligence, surveillance, and reconnaissance), and the capability advantages that they confer are low-profile or even effectively invisible. These factors place a premium on detailed analysis of the interactions between the Asia-Pacific powers' military capabilities.

While there is much room for discussion over whether or not Asia-Pacific military developments in the early twenty-first century should be classed as an arms race, and for debate about the meaning of the term "arms race" in the contemporary region, it does seem clear that governments and nongovernmental experts alike are increasingly concerned that military competition among powers with security interests in the Asia-Pacific has the potential to undermine further regional security and stability. Given the likely difficulty—if not impossibility, in the foreseeable future—of devising, agreeing on, and implementing arms control measures that would restrain governments in the region from developing, purchasing, or deploying certain types of military equipment, or imposing numerical limitations, the governments in the region are looking for ways of mitigating the dangers posed by military competition.

Governments and experts in the Asia-Pacific have engaged extensively through a variety of regional institutions and processes—such as the ASEAN Regional Forum and the Council for Security Cooperation in the Asia-Pacific—with the intent of enhancing the transparency of military developments; but these efforts have been largely ineffective. See chapter 11 for further discussion on this point. In these circumstances, as Desmond Ball has argued, there is an urgent need for governments and nongovernment analysts to think constructively about other mechanisms that might help to reduce the dangers of regional military competition. A good starting point might be efforts to control the risks ensuing from increasingly capable armed forces in the region coming into contact and, potentially, accidental conflict with each other. In the maritime sphere, such potential constraints might help to reduce the risk of escalation in arenas of strategic competition, such as the East China Sea and South China Sea. A useful initial step in this direction was the agreement in April 2014 by twenty-one navies—including those of China, Japan, and the United States—to adopt a Code for Unplanned Encounters at Sea that would, among other measures, prohibit the "locking-on" of weapon-guidance radars.[64] If such measures, however, are to be useful in the context of intensifying military competition in the Asia-Pacific, they will need to be seen as complementary to more comprehensive agreements aimed at avoiding conflict, such as the Code of Conduct for parties to the South China Sea disputes, on which China and members of the Association of Southeast Asian Nations began consultations during 2013.

Key Points

- The arms race concept remains highly contested, though action/reaction dynamics remain a key aspect of most definitions.
- Scholars also remain divided regarding whether the primary causes of arms races are external or internal factors. They also disagree over whether arms races are inherently irrational and destabilizing, or whether they are indicative of rational state behavior that is a source of stability.
- Many Asia-Pacific countries are modernizing their armed forces in response to pervasive regional insecurity; meanwhile, rapid economic growth has provided governments with the resources to fund extensive military modernization programs.
- Consistent with long-standing debates about the arms race concept, there is also considerable disagreement over whether Asian military modernization is consistent with arms-racing behavior.
- Whether or not arms-racing is occurring in the Asia-Pacific, governments there are increasingly looking for ways of mitigating the dangers posed by growing military competition.

Questions

1. How do arms races differ from the military modernization processes in which many nation-states routinely engage?

2. How and why are countries in the Asia-Pacific investing in their armed forces?

3. Why do scholars reach such different conclusions regarding whether an arms race is occurring in the Asia-Pacific?

4. Would arms-racing behavior in the Asia-Pacific be more likely to have a stabilizing or a destabilizing effect on regional security?

5. What do you see as the best way to describe the military development programs occurring in the Asia-Pacific?

Guide to Further Reading

Ball, Desmond. "Arms and Affluence: Military Acquisitions in the Asia-Pacific Region." *International Security* 18, no. 3 (Winter 1993): 78–112. http://dx.doi.org/10.2307 /2539206.
This classic article, written by one of the region's most influential academic strategists, examines how Asian states developed their armed forces in the early 1990s.

Buzan, Barry, and Eric Herring. *The Arms Dynamic in World Politics.* Boulder, CO: Lynne Reinner, 1998.
This is a useful introduction to the arms race concept. It is an oft-cited book that provides a particularly helpful overview of the various possible causes (both external and internal) and consequences of arms races.

Huntington, Samuel P. "Arms Races: Prerequisites and Results." *Public Policy* 8, no. 1 (1958): 41–86.
This is one of the earliest and still most influential pieces of scholarship analyzing the arms race concept. A central feature of arms-racing behavior identified in this article is its quality of "reciprocal interaction," meaning that two states (or coalitions thereof) are consciously increasing their weaponry against the backdrop of a disagreement as to the proper balance of military power between them.

International Institute for Strategic Studies. *The Military Balance 2016.* Abingdon, UK: Routledge for International Institute for Strategic Studies, 2016.
This is a detailed, up-to-date survey of the organization and capabilities of armed forces worldwide, as well as national defense budgets. Chapter 6 focuses on Asian countries' defense spending and armed forces.

Till, Geoffrey. *Asia's Naval Expansion: An Arms Race in the Making?* Adelphi Series 52, nos. 432–33 (2012).
Written by a leading naval historian and scholar of maritime strategy, this book examines naval competition and cooperation in the Asia-Pacific, arguing that though an arms race is not yet occurring, there is risk one will emerge.

Notes

1. See, e.g., Colin S. Gray, "Arms Races and Other Pathetic Fallacies: A Case for Deconstruction," *Review of International Studies* 22, issue 3 (July 2006): 323–35.

2. For further reading, see Paul Kennedy, *The Rise of Anglo-German Antagonism, 1860–1914* (London: Ashfield Press, 1987).

3. Samuel P. Huntington, "Arms Races: Prerequisites and Results," *Public Policy* 8, no. 1 (1958): 41.

4. Colin S. Gray, "The Arms Race Phenomenon," *World Politics* 24, no. 1 (October 1971): 40.

5. Huntington, "Arms Races," 42.

6. Barry Buzan and Eric Herring, *The Arms Dynamic in World Politics* (Boulder, CO: Lynne Reinner, 1998), 80.

7. For further reading on the external causes of arms races, see Charles L. Glaser, "The Causes and Consequences of Arms Races," *Annual Review of Political Science* 3 (June 2000): 253–56.

8. For further reading on the internal causes of arms racing behaviour, see Buzan and Herring, *Arms Dynamic*, 101–18.

9. Bruno Tertrais, "Do Arms Races Matter?" *Washington Quarterly* 24, no. 4 (Autumn 2001): 125.

10. Huntington, "Arms Races," 63.

11. Alexander Wendt, "Identity and Structural Change in International Politics," in *The Return of Culture and Identity in IR Theory*, ed. Yusef Lapid and Friedrich Kratochwil (Boulder, CO: Lynne Rienner, 1996), 50.

12. International Institute for Strategic Studies, *The Military Balance 2015* (London: Routledge for International Institute for Strategic Studies, 2015), 33.

13. "First Plenary Session: the US Approach to Regional Security, Twelfth Asia Security Summit, Singapore, May 31–June 2, 2013," in *The Shangri-La Dialogue* (London: International Institute for Strategic Studies, 2013), 18–21.

14. Ibid.

15. David Gompert and Terrence Kelly, "Escalation Cause: How the Pentagon's New Strategy Could Trigger War with China," *China-US Focus*, August 8, 2013, http://www.chinafocus.com/peace-security/escalation-cause-how-the-pentagons-new-strategy-could-trigger-war-with-china.

16. Information Office of the State Council of People's Republic of China, "The Diversified Employment of China's Armed Forces," April 2013, http://news.xinhuanet.com/english/china/2013-04/16/c_132312681.htm.

17. Edward Wong and Chris Buckley, "China's Military Budget Increasing 10 percent for 2015," *New York Times*, March 4, 2015, http://www.nytimes.com/2015/03/05/world/asia/chinas-military-budget-increasing-10-for-2015-official-says.html?_r=0.

18. Dave Majumbar, "China's First Carrier Conducts Flight Trials with J-15 Flying Shark Fighters," *Flightglobal*, November 25, 2012; "Xi Boards Liaoning Aircraft Carrier, Watches Training," Xinhuanet, August 30, 2013, http://news.xinhuanet.com/english/china/2013-08/30/c_132678449.htm.

19. "Work Under Way on China's Second Aircraft Carrier at Dalian Yard," *South China Morning Post*, January 19, 2014, http://www.scmp.com/news/china/article/1408728/work-under-way-chinas-second-aircraft-carrier-dalian-yard.

20. J. Michael Cole, "New Chinese Ship Causes Alarm," *Taipei Times*, May 31, 2012, http://www.taipeitimes.com/News/front/archives/2012/05/31/2003534139.

21. Wendell Minnick, "China Pursues Systems to Keep US Forces at Bay," *Defense News*, September 17, 2013, http://www.defensenews.com/article/20130917/DEFREG03/309160021/; "China Developing 'World's Most Accurate' Cruise Missiles," *Want China Times*, February 25, 2014.

22. Timothy Farnsworth, "China Conducts Missile Defense Test," *Arms Control Today*, March 2013, http://www.armscontrol.org/act/2013_03/China-Conducts-Missile-Defense-Test.

23. Andrea Shalal-Esa, "US Sees China Launch as Test of Anti-Satellite Muscle: Source,"

Reuters, May 15, 2013, http://www.reuters.com/article/2013/05/15/us-china-launch -idUSBRE94E07D20130515.

24. International Institute for Strategic Studies, *The Military Balance 2014* (Abingdon, UK: Routledge for International Institute for Strategic Studies, 2014), 204.

25. "Panel to Delay Report on Collective Self-Defense until May," *Japan Times*, April 2, 2014, http://www.japantimes.co.jp/news/2014/04/02/national/panel-to-delay-report -on-collective-self-defense-until-may/#.Uzy4r032Ndg.

26. International Institute for Strategic Studies, *Military Balance 2014*, 201.

27. Kiyoshi Takenaka and Nobuhiro Kubo, "Japan Relaxes Arms Export Ban to Fortify Defense," Reuters, April 1, 2014, http://uk.reuters.com/article/2014/04/01/uk-japan -defense-idUKBREA2U1VO20140401.

28. Martin Fackler, "Shinzo Abe Has Eyes on Revising Constitution in Japan," *New York Times*, December 24, 2014, http://www.nytimes.com/2014/12/25/world/asia/japan -shinzo-abe-begins-new-term-with-push-to-revise-constitution.html.

29. Santanu Choudhury, "India Increases Military Budget by 11 percent to Nearly $40 billion," *Wall Street Journal*, February 28, 2015, http://www.wsj.com/articles/india -increases-military-budget-by-11-to-nearly-40-billion-1425124095.

30. Vivek Raghuvanshi, "Indian Army to Upgrade T-90 Tanks with Domestic Help," *Defense News*, March 10, 2014, http://www.defensenews.com/article/20140310/DEFREG03 /303100026/Indian-Army-Upgrade-T-90-Tanks-Domestic-Help.

31. "VSHORAD—India's Next Big Air Defense Program," *Defense Update*, April 1, 2012, http://defense-update.com/20120401_vshorad-indias-next-big-air-defense-program .html.

32. "Eurocopter Hopeful with India," *India Strategic*, March 2013, http://www .indiastrategic.in/topstories1961_Eurocopter_hopeful_India.htm.

33. "PAK-FA/FGFA/T50: India, Russia Cooperate on 5th-Gen Fighter," *Defense Industry Daily*, February 24, 2014; V. Narayana Murthi, "Advanced Medium Combat Aircraft by 2018," *New Indian Express*, January 7, 2014, http://www.newindianexpress.com/states /tamil_nadu/Advanced-Medium-Combat-Aircraft-by-2018/2014/01/07/article1987224 .ece.

34. N. C. Bipindra, "N-Powered Aircraft Carrier in the Works," *New Indian Express*, December 6, 2013, http://www.newindianexpress.com/nation/N-powered-Aircraft-Carrier -in-the-Works/2013/12/06/article1930757.ece.

35. "India's Largest Carrier INS *Vikramaditya* arrives at Karwar Home Port," *Defense Update*, January 8, 2014, http://defense-update.com/20140108_vikramaditya_arrives_at _karwar.html#.U0SyqU32Ndg.

36. "India's Nuclear Submarine Projects," *Defense Industry Daily*, January 21, 2014.

37. Josy Joseph, "Scorpene Submarine Project to Miss Target Again as Spanish Consultants Quit," *Times of India*, April 15, 2013, http://timesofindia.indiatimes.com /india/Scorpene-submarine-project-to-miss-target-again-as-Spanish-consultants-quit /articleshow/19551830.cms.

38. "Indian Navy Receives Second P8I Maritime Patrol Aircraft," *The Hindu*, November 16, 2013, http://www.thehindu.com/news/national/indian-navy-receives-second -p8i-maritime-patrol-aircraft/article5357776.ece; "First Indian MiG-29 Fighter Jet Lands

on *Vikramaditya*," *RIA Novosti*, February 7, 2014, http://en.ria.ru/military_news/20140207/187307197/First-Indian-MiG-29-Fighter-Jet-Lands-on-Vikramaditya.html.

39. International Institute for Strategic Studies, *The Military Balance 2015* (Abingdon, UK: Routledge for International Institute for Strategic Studies, 2015), 164.

40. Ibid., 159.

41. Franz-Stefan Grady, "What to Expect from Russia's Pacific Fleet in 2015," *The Diplomat*, March 2015.

42. Bernard D. Cole, *Asian Maritime Strategies: Navigating Troubled Waters* (Annapolis, MD: Naval Institute Press, 2014), 83–85.

43. Thomas Grove, "Russia Shows Off Military Might as Budget Gets Squeezed," *Wall Street Journal*, August 27, 2015.

44. Australian Government, *Defense Economic Trends 2014* (Canberra: Australian Government, 2014), 14.

45. "Submarines for Indonesia," *Defense Industry Daily*, February 18, 2014.

46. "Rolling Sales: Indonesia Becomes the Latest Buyer of German Tanks," *Defense Industry Daily*, November 17, 2013.

47. "Indonesia: AH-64D APACHE Block III LONGBOW Attack Helicopters," news release, Defense Security Cooperation Agency, September 21, 2012, http://www.dsca.mil/major-arms-sales/indonesia-ah-64d-apache-block-iii-longbow-attack-helicopters.

48. International Institute for Strategic Studies, *Military Balance 2015*, 293.

49. "Vietnam Holds Flag-Raising Ceremony for *Kilo*-Class Subs," *Tuoinews.vn*, April 4, 2014, http://tuoitrenews.vn/society/18810/vietnam-holds-flagraising-ceremony-for-kiloclass-subs.

50. Ridzwan Rahmat, "Damen Schelde Reveals Design of New Vietnamese *Sigma*-Class Corvettes," *IHS Jane's Defense Weekly*, March 6, 2014.

51. "Vietnam's Russian Restocking: More Frigates Ordered," *Defense Industry Daily*, March 6, 2014.

52. "Vietnam to Buy Dozen Sukhoi Fighter Planes from Russia," *Thanh Nien News*, August 23, 2013, http://www.thanhniennews.com/politics/vietnam-to-buy-dozen-sukhoi-fighter-planes-from-russia-1428.html.

53. Desmond Ball, "Arms and Affluence: Military Acquisitions in the Asia-Pacific Region," *International Security* 18, no. 3 (Winter 1993–94): 78–112.

54. Ibid.

55. Desmond Ball, "Northeast Asia: Tensions and Action-Reaction Dynamics," paper prepared for International Institute for Strategic Studies workshop "Sub-Regional Dynamics and Regional Order," Singapore, November 3, 2011.

56. Ibid.

57. Ibid.

58. Ibid.

59 Robert Hartfiel and Brian L. Job, "Raising the Risks of War: Defense Spending Trends and Competitive Arms Processes in East Asia," *Pacific Review* 20, no. 1 (March 2007): 2.

60. Geoffrey Till, *Asia's Naval Expansion: An Arms Race in the Making?* Adelphi Series 52, nos. 432–33 (2012): 18.

61. Ibid., 239.

62. Christian Le Miere, "The Spectre of an Asian Arms Race," *Survival* 56, no. 1 (February–March 2014): 139–56.

63. Richard Bitzinger, "A New Arms Race? Explaining Recent Southeast Asian Military Acquisitions," *Contemporary Southeast Asia* 32, no. 1 (April 2010): 50–69.

64. "21 Navies Support Ban on Radar-Lock," *Japan News*, April 21, 2014.

7 Maritime Security—Will Asia's Next War Occur at Sea?

James Manicom

Reader's Guide

This chapter considers the question of maritime security issues in the Asia-Pacific. It examines the key straits and shipping channels, as well as debates over economic zones, fishing, and codes of conduct at sea. It explains why maritime space has become more important to states in the region and examines the key stakes at sea. As a case study, it uses the current disputes in the South China Sea with the overlapping claims of authority, claims to resources, and alliance relationships as evidence of the challenges the region faces. It explores the following key questions: Why is maritime security important in the region? What are the primary maritime security flashpoints in the region? How does energy and resource scarcity affect maritime and regional security?

Introduction

In many ways maritime security issues in the Asia-Pacific region were a catalyst for the shift in US policy between 2009 and 2011. This "pivot" or "rebalance" consists of commitments designed to reassure the region that the United States intends to remain the preeminent power. Although not formally announced until November 2011, Barack Obama campaigned as a Pacific president, and in 2009 he sent an early political signal of his priorities by ratifying the Treaty of Amity and Cooperation of the **Association of Southeast Asian Nations (ASEAN)**, which is a prerequisite for joining the region's premier regional economic and security institutions, respectively, the **East Asia Summit** and the **ASEAN Defense Ministers' Meeting Plus**. However, the diplomatic effort began in earnest in 2011 following a number of confrontations at sea between military and civilian vessels from countries across the region. The subsequent deterioration of political relationships in the region was perceived by US policymakers as posing a threat to the key US regional interests—regional stability and safe and secure navigation through the region's waters. By virtue of its geography, the Asia-Pacific is an inherently maritime region, which places maritime issues at the forefront of regional security considerations.

The Asia-Pacific is a maritime region of oceans, seas, and archipelagoes. The littoral seas, such as the South China Sea and East China Sea, are "semi-enclosed" seas bordered on several sides by different states. Furthermore, the region is also home to a number of narrow straits—such as the Malacca Strait, the Lombok Strait, and the Makassar Strait—which have all become important thoroughfares for regional trade. The sea provides people and their governments with a source of income and a conduit for trade. Furthermore, every nation in the region has been subjected to some kind of trauma from the sea, whether it was Commodore Perry's Black Ships arriving in Tokyo Bay to force open Japanese markets (1852), treaty ports established by the European powers along Qing China's coast (1840s–1940s), or the outright colonization of much of Southeast Asia (1511–1984).[1] The sea is thus a paramount security issue for countries in the region; it touches on the livelihoods of its people, the security of food and energy supplies, the stability of sea lanes that bring its goods to global markets, and above all the national sovereignty that each government protects so dearly. Moreover, as a result of climate change and affected weather patterns, from time to time the sea itself is capable of threatening everything on land.

This chapter begins by outlining the maritime geography of the Asia-Pacific region and the salient legal issues that govern the oceans. It then surveys the stakes that countries in the region have at sea and outlines threats to these stakes. The final section looks at competing claims to the islands, rocks, and atolls of the South China Sea to illustrate the myriad competing dynamics in maritime East Asia. Although US strategy in the post–Cold War era has always been intended to strengthen regional security by deterring states' future adventurism, the strategic shift announced in 2011 was notable in its overt effort to clearly articulate a sustained US commitment to the region, including the Trans-Pacific Partnership regional trade agreement and the long-term shift of military power to the Asia-Pacific.

Maritime Geography in the Asia-Pacific

Overlapping maritime boundary disputes are a product of the geographical makeup of East Asia and the widespread ratification of the **United Nations Convention on the Law of the Sea (UNCLOS)**. UNCLOS allows states to make claims to maritime space from their coastlines. States have different rights within these zones that are balanced against the rights of the rest of the world, which uses the ocean for navigation and overflight. Coastal state **jurisdiction** is strongest within the territorial sea that stretches the 12 nautical miles (nm) offshore. States have basically the same rights here as they do over their own land, with one exception. Ships from other countries are allowed to pass through the territorial sea "innocently"; they cannot stop, military ships must fly their flag, and submarines must travel on the surface. Beyond the territorial sea lies the contiguous zone, an additional 12 nm zone in which coastal state customs, immigration, and sanitation laws apply. The third type of zone, the **Exclusive Economic Zone (EEZ)**, stretches up to 200 nm from the coast. Herein coastal state jurisdiction is limited even further, simply to economic rights over fish and resources on and beneath the seabed. Coastal states also retain jurisdiction over marine scientific research, artificial islands, and structures, as well as responsibility for the management of resources within this zone. Problematically, because of the close proximity of the East Asian states to one another, most of these claimed zones overlap, raising the question of which nation gets to do what, where, and which one gets the profit. Furthermore, regional states such as Japan, China, South Korea, Vietnam, and Malaysia have claimed jurisdiction over their continental shelf beyond 200 nm. These are called "extended continental shelves," which give the coastal state authority over the seabed and the subsoil, but not the water above.

It is important to note that in addition to creating entitlements to maritime space, UNCLOS also creates a number of obligations for coastal states. These include having consideration for the interests of other states that use the ocean, conservation and stewardship of the marine environment, and the obligation to cooperate and settle disputes peacefully. Furthermore, UNCLOS outlines a number of different procedures that states can use to cooperate, and it created its own court to settle disputes. Asia-Pacific states have generally been very good at using these mechanisms. Bangladesh has twice settled maritime boundary disputes in the Bay of Bengal, with Burma/Myanmar and India, using the UNCLOS tribunal. Outside UNCLOS, the International Court of Justice has occasionally settled sovereignty disputes and drawn maritime boundaries. Malaysia has twice used the Court to help settle disputes with Indonesia and Singapore. Regrettably, few parties have shown an interest in using these mechanisms in the East China Sea and the South China Sea, with the exception of the Philippines. In 2012 Manila initiated a special UNCLOS proceeding to challenge China's claims to the South China Sea. Beijing has refused to participate in the process on the grounds that the tribunal does not have the authority to hear the Philippines' complaint.

Further complicating the situation is a host of disputed claims to **sovereignty** over islands, rocks, and reefs: In Northeast Asia, Japan and South Korea both claim the Dokdo/Takeshima rocks; China and Japan both claim the Senkaku/Diaoyu Islands; and Japan and Russia each claim the Northern Territories / Southern Kuriles Islands (table 7.1). These **territorial disputes** are a function of poorly made claims to sovereignty and oversights in the San Francisco Peace Treaty that ended World War II in Asia.[2] The Dokdo/Takeshima Islands were seized by Japan in 1905, on the eve of its colonization of Korea. South Korea retook the islands in 1954, while Japan was demilitarized, and the issue has festered ever since. Japan claimed the Senkaku/Diaoyu Islands in 1895 on the basis that they were unoccu-

Table 7.1 Disputed Territories in the East China Sea and the South China Sea

Territory Name	Claimed by
Senkaku/Diaoyu/Dokdo	Japan, China, South Korea
Paracel	China, Taiwan, Vietnam
Spratly	Brunei, China, Malaysia, Philippines, Taiwan, Vietnam
Pratas	China, Taiwan
Macclesfield	China, Taiwan
Scarborough Shoal	China, Taiwan, Philippines
Natuna	Indonesia[a]

[a]The islands themselves are only claimed by Indonesia. Their waters may be claimed by China.

pied, although China maintains it had used them as navigational aids since the Ming Dynasty. China, Taiwan, and Japan all reissued claims after the discovery of promising oil and natural gas reserves in 1969. Russia captured the Kuriles Islands in the closing days of World War II, and Japan surrendered its claim to them under the San Francisco Peace Treaty. However, Russia never signed the treaty, and the issue has endured because Japan views the Northern Territories / Southern Kuriles as separate from the rest of the Kuriles Islands, which it accepts as Russian.

Although these rocks and islands are for the most part small and uninhabitable, their economic value is enhanced by the fact that they may or may not be entitled to the maritime zones listed above. Furthermore, the politics of territorial issues is exacerbated by the negative images many governments and peoples hold of their territorial rivals as a result of unsettled historical grievances related to perceived injustices suffered at the hands of other states.[3] These disputes thus link realist concerns about sovereignty and economics with strong constructivist themes of identity, historical memory, and norms. It is impossible to separate these material and nonmaterial issues and to say definitively that one is more important in policymakers' minds, or that one preceded or led to the other. Public opinion in these countries, especially **nationalism** in East Asia, has hardened state postures, has prevented accommodation between claimant states, and has even been a source of escalation. There is no better demonstration of this than the efforts of the conservative Japanese politician Ishihara Shintaro to provoke a crisis with China over the Senkaku/Diaoyu Islands by attempting to buy them on behalf of the City of Tokyo in 2012.

These disputes encourage states to allocate resources to the military—particularly the navy and air force—as well as toward coastal enforcement vessels. Many East Asian states have fielded large numbers of modern surface ships, complete with advanced air defense capabilities and technologically advanced war-fighting capabilities, including land- and sea-based cruise missiles.[4] Most troubling has been the dramatic rise in the number of advanced diesel-powered submarines, particularly by South China Sea claimants like Vietnam and Malaysia; these submarines are quieter and harder to detect than ever before. Although more coast guard vessels being deployed is preferable to more military vessels being deployed, it needs to be noted that the bulk of the tensions at sea in the region have occurred when civilian coast guards enforce maritime jurisdiction in contested areas. This includes the standoff between China and the Philippines at Scarborough Shoal in April 2012 and a number of confrontations between Chinese and Vietnamese vessels near the Paracel Islands.

The proliferation of armed government ships at sea is problematic for a number of reasons. First, maritime jurisdiction within East Asia is contested, so all parties exercise authority in areas where other states also claim the same right. Coastal states also differ over the degree of authority that can be exercised in coastal waters, which has been illustrated in a number of confrontations between Chinese and American ships off the coast of China.[5] Second, there is little transparency between regional navies or coast guards, which raises the risk of escalation. It is only a matter of time before a maritime accident turns deadly, which could escalate to a shooting war between two or more countries in the region. Tragically, the role of these agencies in maritime boundary disputes distracts from the potentially important role that coast guard cooperation can play in improving political relations while addressing urgent security issues like piracy and human smuggling.[6] **Maritime boundary disputes** are exacerbated by the growing material importance of income from the ocean for countries' national development goals.

Maritime Stakes in a Maritime Region

Since UNCLOS came into force in 1994, all the East Asian states have become more interested in maritime areas. Liberalism argues that this is an expected outcome of regimes and institutions as they direct attention and focus toward common challenges. The more optimistic also hope that such frameworks will encourage discussion and enable peaceful dispute resolution. Asia leads the world in fish consumption and production, and in the number of people employed in fisheries industries, including aquaculture. Asia boasts 74 percent of the world's fishing vessels, and China alone is responsible for 34 percent of global fish production.[7] Furthermore, according to the US Department of Energy, the South China Sea contains 190 trillion cubic feet of natural gas.[8] Natural gas consumption is expected to rise because it plays an important role in a region afflicted with an acute sense of energy insecurity as states seek to diversify the locations and types of energy they import.[9]

As noted above, the Asia-Pacific countries have historically had a decidedly mixed experience with the ocean; it is simultaneously a source of security and insecurity. The region suffers from acute shortages that exacerbate energy and food insecurity, climate catastrophes, and humanitarian disasters.[10] And the region's countries are consuming increasingly vast amounts of energy every year, which, due to the region's relative poverty in primary energy sources, is met with imported supplies. The region consumes 39 percent of global energy, but has less than 3 percent of global oil resources and 8 percent of global natural gas resources.[11] The pursuit of **energy security**, defined as access to sufficient energy supplies at affordable prices, is thus an overriding regional prerogative that is perceived to be linked to economic growth.[12] There are a number of threats to Asian energy security supplies that stem from both state and nonstate factors. State-based threats include situations in which a state or a company somewhere along the energy supply chain deliberately attempts to disrupt the flow of primary energy sources like coal, natural gas, uranium, or oil. These include politically motivated market manipulation by supplier states, the naval blockade of sea lanes, or hoarding of supplies. Nonstate threats have proven more frequent and include disruption along the supply chain due to terrorist attacks or piracy, natural disasters, and demand fluctuations in energy-importing states that raise prices elsewhere.

The prospect of disruption of the sea lanes is of particular concern for the export-oriented economies of the Asia-Pacific and is tied to their perceptions of national security. More than 90 percent of global trade travels by ship. According to Commander Admiral Samuel Locklear III, head of the US Pacific Command, 50 percent of global container traffic and 70 percent of ship-borne energy passes through this region.[13] Almost half of China's gross domestic product travels by ship, and 80 percent of its imported oil travels through the narrow Malacca Strait. The security of **sea lanes of communication** played an important role in loosening the restrictions on Japan's Navy because of Japan's near-exclusive reliance on imported energy and the export-oriented nature of its economy. As early as 1978, Tokyo reinterpreted its security agenda to include patrols of its waters 1,000 nm from its shore. Twenty years later, as China's maritime interests expanded with its growing reliance on maritime-based trade, Chinese leaders spoke of a "Malacca dilemma" reflecting anxieties about the security of the Malacca Strait. It is thus unsurprising that maritime security is a chief concern for powers in the region.

Among these threats to maritime security, **piracy** has been the source of the most coherent exercise of interstate cooperation. Following a steady rise in pirate attacks in the Malacca and Singapore straits in the late 1990s and early 2000s, pressure grew on the littoral states of Malaysia, Singapore, and Indonesia to address the issue. Confronted by the prospect of US Navy patrols in their territorial waters, the three countries worked with Japan to establish a piracy-reporting center and improve regional maritime domain awareness. Since its inception, the result has been a dramatic drop in incidents of piracy and armed robbery at sea. In general terms, maritime security has long been a source of positive interaction and capacity building between ASEAN states and their dialogue partners because so many maritime security issues pose common threats to all, including pollution control, maritime safety, building navigation infrastructure, and coast guard capacity building. Specifically, maritime security is one of the most consistently studied areas in regional track-two dialogues. At the official level, both the **ASEAN Regional Forum** and the ASEAN Defense Ministers' Meeting Plus have working groups on maritime security.

Maritime security is also vital for humanitarian crises in the region. There has been an increase in extreme weather events across the Asia-Pacific during the past two decades. The frequency of meteorological disasters in the period 2001–10 increased 66 percent over the previous decade.[14] The fact that East Asian populations, and, by extension, their infrastructure, are concentrated on coastlines increases the likelihood that they will be damaged by the storms that afflict the region. As a product of industrialization and urbanization, the material cost of natural disasters in China more than doubled in the period between 1995 and 2004 compared with the ten years prior. It increased tenfold in Japan and ninefold in South Korea.[15] In addition to high death tolls and the costs of reconstruction, these events can also increase the number of people displaced, which in turn has implications for nearby cities and countries and the region as a whole. Low-lying areas in states like Bangladesh and the Philippines are particularly vulnerable, although most East Asian cities are near the coast. It is thus unsurprising that Admiral Locklear has described climate-related disruptions as the most probable security challenge in East Asia.[16] As illustrated by the international response to Typhoon Haiyan in 2013 and to the Asian tsunami in 2004, relief from these disasters often comes via the sea. Naval ships from regional states—including Australia, China, Japan, and the United States—have all contributed to humanitarian assistance and disaster relief operations. Although not unique to Asia, ship-borne delivery of international assistance is a compelling reminder that despite being instruments of war, navies are simultaneously tools of public **diplomacy**.

These maritime stakes are an excellent illustration of the differing perspectives outlined in this book's introduction. For those that identify the state as the primary security referent—and this includes several governments in the region—the future of the region will hinge on the management of security threats identified as threats to the state, including the resources required for its material well-being. Security, in this view, is achieved by the application of military power to secure access or to deny access to others. Curiously, this perspective lends itself to an optimistic prognosis for the region—military power can be deterred, and material stakes of disputed space can be divided.

By contrast, those that see a state's identity as the security referent should be much more pessimistic. However important fish and energy resources are, these pale in comparison with the intangible value that peoples and governments place on their sense of self when this hinges on the views of another. In this context, the history between claimants to disputed territory is important. For instance, the world is far more concerned about the territorial dispute between China and Japan than that between Canada and Denmark (over Hans Island). In this context, security is achieved not by securing access to material stakes but by being able to subject a rival state to one's will. This is a far more dangerous set of circumstances because a mutually satisfactory outcome is difficult to identify.

Finally, those that focus on the human being as the security referent are the most pessimistic of all. As long as states argue over disputed maritime territory, they are also party to the collective human failure to respond to the tremendous pressures human activity has placed on the maritime ecosystem in the Asia-Pacific. Collective action problems that affect human well-being such as regional fisheries management, regional energy infrastructure, and transnational crime prevention—including piracy—are all hostage to the deterioration of the web of political relationships across the region.

National Sovereignty: The Most Important Maritime Security Issue?

Despite the number of security issues that threaten the Asia-Pacific as a whole, the region regrettably remains preoccupied with narrow territorial and jurisdictional concerns. In the maritime domain at least, zero-sum realist conceptions of world politics tends to be the norm. The tragedy of this situation is twofold. First, disputed rocks and islands in the East China Sea, the South China Sea, and the Sea of Japan hold little intrinsic value, yet claimant states are prepared to grandstand for the sake of national pride and domestic politics. This situation has led to a number of risky encounters at sea, and it means that the threat of war looms large over East Asia. The tragedy's second dimension is that preoccupation with narrow territorial interests detracts from meaningful cooperation on the myriad maritime security challenges that present a common threat to all the East Asian peoples.

For instance, China and Japan have engaged in military posturing in the East China Sea on a number of occasions. These incidents have occurred in the vicinity of the Senkaku/Diaoyu Islands, near the Chinese-run Chunxiao natural gas field and in the maritime approaches that lead to the Pacific Ocean, such as the Miyako Strait. These incidents have included confrontations between coast guard vessels conducting marine scientific research (March and September 2010), close encounters between Chinese coast guard helicopters and Japanese destroyers (March 2011), and the locking of fire control radar from a Chinese navy ship onto a Japanese one (January 2013). Any one of these incidents could have triggered a crisis, which, due to poor communication between ships and the domestic costs of conciliation, could escalate to war. Moreover, the poor state of the China-Japan relationship has impeded cooperation on regional fisheries management, improved environmental standards, global currency reform, and regional trade promotion.

The most complex territorial dispute in the region is the sovereignty dispute over the Spratly Islands in the South China Sea and the maritime boundary dispute over the surrounding sea areas. The South China Sea dispute includes six claimants that each occupy some number of features, which in many cases are little more than rocks or reefs that are below water at high tide. Claimant states have often constructed various kinds of installations on these features, and in many cases have reinforced the features with concrete to make them nominally inhabitable for the few unfortunate military personnel who stand duty. For instance, in 1999 the Philippines ran the Sierra Madre (a World War II–era American Navy ship) aground on Ayungin Shoal and stationed a small garrison to defend the feature. The troops are resupplied by airlift. The largest of the disputed features, occupied by Taiwan, is Itu Aba, which boasts its own freshwater supply and a 1,200-meter-long runway that could accommodate military aircraft.

For many years the South China Sea dispute was an occasional irritant in regional relations. For instance, in 1995, the Philippines discovered that China had constructed a small structure on Philippine-claimed Mischief Reef. Despite appealing to ASEAN and other regional institutions, the Philippines was unable to change the fact of China's occupation, which likely caused the rash of occupations of features in the late 1990s. Despite the recent spate of confrontations at sea, there is no evidence that any claimant is considering using force to expel the occupying forces of another claimant from an occupied feature. Nevertheless, the risk of conflict should not be underestimated. In 1988 a naval skirmish near Johnson Reef between Chinese and Vietnamese navies resulted in more than seventy Vietnamese casualties.

Political tensions are also exacerbated when claimant countries try to enforce their maritime jurisdiction over rival claimants in areas that both parties claim. According to some reports, in 2011 Chinese navy and coast guard vessels bullied Philippine fishing boats near Palawan Island and harassed marine survey ships in the area. This set the stage for a dramatic confrontation between the Chinese coast guard and a Philippine navy ship near Scarborough Shoal in April 2012. In this case the Philippine ship withdrew under the terms of a mutually agreed-on stand-down in the face of oncoming poor weather. The Chinese ships chose to stay on station and cut off access to Scarborough Shoal following the Philippine withdrawal. And the Chinese ships were also particularly assertive in the enforcement of their annual fishing ban near the Paracel Islands in the northern part of the South China Sea, which included training guns on Vietnamese fishing vessels. This was followed by allegations that the Chinese ships had interfered with a seismic survey ship working for the Vietnamese government by cutting the cable that towed its seismic equipment.[17] China has not been the only aggressor, however. In May 2013 the Philippine coast guard shot and killed a sixty-five-year-old Taiwanese fisherman in disputed waters in the South China Sea, which triggered a four-month deterioration of relations, including economic sanctions, but which ended in a Philippine apology. In both cases the stronger party, China or Taiwan, was able to coerce the smaller state. Unsurprisingly, both incidents led the Philippines to deepen its military ties with its treaty ally, the United States.

As a result of these tensions, regional states have been trying to create conflict avoidance mechanisms. ASEAN and China concluded a Declaration on the Conduct of the Parties of the South China Sea Dispute in 2002, following a decade of steadily rising tensions. The declaration is a nonbinding commitment among the parties to avoid escalating the dispute by occupying new features or doing anything else to exercise jurisdiction over the disputed area. China's acquiescence to such an arrangement needs to be understood in the context of its "Smile Offensive" toward Southeast Asia, which sought to capitalize on the goodwill created by China's constructive policies during the 1997–98 Asian financial crisis and on the United States' preoccupation with regional terrorist groups after the September 11, 2001, terrorist attacks and the 2002 Bali bombing. While the United States was preoccupied with the global war on terror, China was busy building confidence in the region and negotiating trade deals in Southeast Asia. This perception of the United States' distraction—and, following the 2008–9 global financial crisis, of its weakness—is what many attribute to China's growing assertiveness in maritime East Asia.[18] Although the pivot is a long-term strategy to embed the United States into the economic, strategic, and political fabric of the Asia-Pacific region, it was ultimately triggered by growing instability at sea. These issues are at their starkest in the South China Sea. The multilateral institutional framework of Asia, one welcomed and supported by advocates of liberalism in particular, is also considered to face a test of capacity and credibility over this issue.

The South China Sea is a potentially dangerous flashpoint for three reasons. First, escalation could trigger a wider conflict. The Philippines, one of the materially weakest claimants yet one with important nationalist ties to the disputed islets, is a US ally. This means that, if the Philippines were attacked by another claimant as a function of these tensions, the United States would be obligated to defend the Philippines, which in turn could trigger a great power war if the aggressor were China. Indeed, in the summer of 2011 the Philippines sought, and received, a reiteration of the US security guarantee three months before Obama announced the pivot. Beyond its obligations to its lone treaty ally that is a party to the dispute, the United States has long maintained that its interests are the security of the region's sea lanes and that the dispute be resolved peacefully. At the 2010 meeting of the ASEAN Regional Forum, US secretary of state Hillary Clinton arguably changed this stance by offering to mediate regional tensions, which China perceived as a deviation from Washington's stated position of neutrality on the issue of disputed sovereignty.[19]

The second reason that the South China Sea disputes are potentially dangerous is because claimant states have several incentives for confrontation and few incentives for cooperation. All claimants accept that the South China Sea is important for material gain—income from the ocean, in the form of oil and natural gas exploitation or fisheries exploitation. The myriad overlapping claims mean that there is little incentive to conserve either resource. In the case of living resources, fish may move into another country's waters, where fishermen may not exercise restraint in harvesting of them, resulting in a race to the bottom for regional fisheries resources. Although oil and natural gas prices are cyclical, most Asia-Pacific states use state-owned companies to conduct offshore exploration, which removes market considerations from the making and staking of claims.

These material incentives for conflict are exacerbated by domestic disincentives for cooperation. Whether as a function of a national narrative of reclaiming lost territorial and maritime rights (China), as part of domestic accusations of being soft on issues of sovereignty (the Philippines), or of being soft on an historical aggressor (Vietnam), political leaders in all countries face serious domestic opposition to cooperation. In 2007 the Vietnamese people took to the streets to protest Chinese actions against Vietnamese oil and fisheries boats. A trilateral joint development agreement between the Philippines, China, and Vietnam was canceled after Philippine opposition politicians criticized the government for bargaining away Philippine territory in exchange for foreign aid. Finally, in China, domestic political actors have found it useful to forcefully assert China's claims as part of a complex system of bureaucratic politics that rewards those that defend China's sovereignty.[20] Claimants to the South China Sea therefore confront both political and economic incentives to be assertive at sea.

The third reason that the South China Sea is dangerous is because the material value of the space increases the rationale for investment in naval and coast guard assets to police disputed space. East Asia's dynamic economic growth is being invested in the tools of conflict. As a consequence, the region has seen numerous instances of confrontations at sea between coast guard vessels and civilian ships. Moreover, when government vessels confront each other, there are often no communication protocols to alleviate tensions in a crisis. The growing asymmetry between the giant China and its medium-sized and small neighbors in the area of coast guard capacity has yielded a few tactical victories for China—for instance, when the Chinese coast guard outlasted its Philippine counterpart in a confrontation near Scarborough Shoal in April 2012. These victories have exacerbated concern in Washington that the region is on a trajectory for conflict over a number of insignificant rocks and islets.

US Policy and the Future of Maritime Security

The Obama administration's pivot is based on the assumption that US military hegemony in the Asia-Pacific since the end of World War II has maintained regional stability, deterred would-be aggressors, and generally alleviated interstate security concerns so as to allow the region's countries to devote resources to economic growth. This idea has clearly been under stress since the end of the Cold War, and challenges to it have accelerated since the global financial crisis caused the world to doubt the United States' capacity and will to maintain its hegemony. The United States' policy has therefore tried to reinforce its interests in regional security at a time of unprecedented global skepticism about its credibility. For instance, the US government continues to assert that it will abide by the security guarantees offered to its treaty allies in the region—Japan, South Korea, Thailand, the Philippines, and Australia. The Obama administration publicly reiterated its commitment to the defense of the Philippines in 2011, and it has stood by its commitment to Japan on multiple occasions.

The military aspect of the strategy is designed to counter these concerns and has amounted to three new initiatives from the United States: the shift of US conventional military power from the Middle East and Europe to the "Indo-Asia-Pacific"; deepening military ties with traditional and new US allies; and finally, encouraging these allies to cooperate among themselves in order to take a greater share of the security burden.[21] The latter two are the most relevant to maritime security in East Asia, although the third is the most significant departure from previous US policy. First, the pivot has consisted of rhetorical statements in support of US allies and frank statements by US policymakers on maritime issues in the region. In his address to the Australian Parliament in November 2011, President Obama stated unequivocally that "reductions in US defense spending will not—I repeat, will not—come at the expense of the Asia-Pacific."[22] Secretary of Defense Leon Panetta noted at the 2012 Shangri-La Dialogue that 60 percent of the US Navy would be stationed in the Pacific by 2020, in a departure from the standard 50/50 split maintained to this point.[23] A similar rebalancing of air force assets was announced in 2013. A redeployment of Marines to Australia soon followed, as did the planned deployment of the littoral combat ship to Singapore. However, it should be noted that there are in fact few new resources being deployed to the Pacific theater; rather, the rebalancing of forces is a story of reductions in Europe and the Middle East, as US forces turn over responsibility for security to local elements, whether in Ramstein or Kabul. Nevertheless, those that dismiss the pivot as merely rhetoric combined with military cuts fail to comprehend the importance of both words and deeds to US deterrent posture.

The second element of the US pivot is deepening ties with traditional and new allies and partners. This includes developing a more permissive set of access arrangements with a broader array of countries—including nontreaty allies such as India, Malaysia, and Vietnam—modeled on arrangements with Singapore, where the littoral combat ship is based and where US aircraft carriers often come into port. Some scholars and policymakers in the United States have argued for a more forceful reinforcement of US alliance relations with existing and new partners, such as Indonesia, through the augmentation of defense ties, including facilitating Jakarta's purchase of diesel-powered submarines, which are not made by the United States.[24] Already, the United States is deepening military ties with India and has inaugurated an annual defense dialogue with Vietnam. Furthermore, Washington has also created dialogues on regional security in Asia with two traditional extraregional allies, Canada and the European Union. The United States now cooperates with more Asia-Pacific countries than at any time in its history.

The third element has been an effort by US policymakers to encourage the spokes of the "hub-and-spokes" San Francisco alliance system to cooperate among themselves, which marks an unprecedented shift in the United States' preferences by avoiding the micromanagement of military affairs among its allies. This has seen a deepening of defense ties and dialogues between Australia, Japan, and partner countries like India. Japan and India have agreed to hold a trilateral military exercise with the United States, which is an unprecedented development, given the respective pacifist and isolationist foreign policy traditions of these two powers. Japanese support of the Philippines and Vietnam has deepened to include the transfer of civilian maritime enforcement capability and training. India and Vietnam are also considering closer ties between their navies. By far the most coherent US effort has been devoted to improving defense ties between Japan and South Korea, which are at odds over Tokyo's treatment of its historical legacy with its neighbors. Japan and South Korea have negotiated agreements on intelligence sharing as well as acquisitions and cross-servicing—collectively, the basic building blocks of a defense relationship—but both have floundered due to their unpopularity in South Korea.

Conclusion

The Obama administration's efforts to shore up its alliances, deepen its defense ties, and bolster American credibility were triggered by a recognition in the highest reaches of America's foreign policy establishment that the United States had fallen behind in the Asia-Pacific region. Although there is considerable commonality of interest between East Asian governments in a secure and stable maritime realm that benefits "the human," the region has become beholden to security threats defined by "the state." It is thus unsurprising that concerns abound East Asia's next conflict will occur at sea.[25] Maritime security issues lie at the very heart of the US pivot. Properly executed, US policy could stabilize the region by deterring reckless behavior and encouraging cooperative action on the myriad human security challenges that confront the region. However, US policy alone can never resolve the differences that divide maritime East Asia. Only the exercise of leadership by the region's countries, particularly China, to genuinely move beyond their historical legacies and overcome the zero-sum perceptions of the value of disputed maritime space will be sufficient to build lasting peace and stability in the region.[26]

Key Points

- Maritime security concerns lie at the heart of the Obama administration's policy toward the Asia-Pacific.
- States place value on maritime space in different ways, which affects the perception of the importance of the ocean.
- UNCLOS is an important factor in maritime security; it has provided the pretext for conflict but also provides pathways to cooperation.
- A strong and credible United States role in the region is important for regional stability; it deters aggression and reassures partners.
- The region's states are agents in their own security; they are increasingly willing to cooperate to increase their **autonomy** from the region's great powers.

Questions

1. Why is maritime security important in the Asia-Pacific? Which theoretical paradigm best explains why this is the case?

2. What are the primary maritime security flashpoints in the Asia-Pacific?

3. Why is the South China Sea seen as a primary security concern?

4. How do energy and resource scarcity affect maritime and regional security?

5. How important are seabed resources likely to be in the future?

6. How does piracy affect maritime security?

Guide to Further Reading

Bateman, Sam, and Ralf Emmers, eds. *Security and International Politics in the South China Sea: Towards a Co-operative Management Regime.* London: Routledge, 2009.
    This book explores the politics of the South China Sea issue, with specific reference to ASEAN-centric efforts to enmesh regional powers in institutional mechanisms focused on conflict avoidance and nontraditional security cooperation.

Bush, Richard. *Perils of Proximity: China-Japan Security Relations.* Washington, DC: Brookings Institution Press, 2010.
    This book explores the institutions of the China-Japan security relationship, with specific reference to maritime issues.

International Crisis Group. "Stirring Up the South China Sea (I)," *Asia Report* no. 223 (April 23, 2012). http://www.crisisgroup.org/~/media/Files/asia/north-east-asia /223-stirring-up-the-south-china-sea-i.pdf
    This landmark report outlines the key actors in Chinese policymaking, with reference to maritime space, specifically the South China Sea.

Manicom, James. *Bridging Troubled Waters: China, Japan, and Maritime Order in the East China Sea.* Washington, DC: Georgetown University Press, 2014.
  This book offers a framework for analyzing how states perceive the value of maritime space and uses it to explain the rise and fall of cooperation between China and Japan in the East China Sea.

Schofield, Ian, ed. *Maritime Energy Resources in Asia: Legal Regimes and Cooperation.* Seattle: National Bureau of Asian Research, 2012.
  This report takes a critical view of the resource wealth of maritime Asia and examines cooperative pathways forward.

Notes

1. The events noted here occurred during the Qing period in China; the People's Republic of China was only founded in 1949.

2. Kimie Hara, *Cold War Frontiers in the Asia-Pacific: Divided Territories in the San Francisco System* (New York: Routledge, 2007).

3. James Manicom, "The Interaction of Material and Ideational Factors in the East China Sea Dispute: Impact on Future Dispute Management," *Global Change, Peace, and Security* 20, no. 3 (2008): 375–97.

4. Richard A. Bitzinger, "A New Arms Race? The Political Economy of Maritime Military Modernization in the Asia-Pacific," *Economics of Peace and Security Journal* 4, no. 2 (2009): 32–37.

5. James Manicom, "Beyond Boundary Disputes: Understanding the Nature of China's Challenge to Maritime East Asia," *Harvard Asia Quarterly* 12, nos. 3–4 (2010): 46–53.

6. Sam Bateman, "Coast Guards: New Forces for Regional Order and Security," *Asia Pacific Issues*, no. 65 (2003).

7. Fisheries and Agriculture Organization, *World Review of Fisheries and Aquaculture* (Geneva: Food and Agriculture Organization, 2012), 3–12.

8. US Energy Information Administration, *Country Analysis Brief: South China Sea* (Washington, DC: US Department of Energy, 2013), 2.

9. Vlado Vivoda and James Manicom. "Oil Import Diversification in Northeast Asia: A Comparison between China and Japan," *Journal of East Asian Studies* 11, no. 2 (2011): 223–54.

10. Alan Dupont, *East Asia Imperilled: Transnational Challenges to Security* (Cambridge: Cambridge University Press, 2001).

11. BP (formerly British Petroleum), *BP Statistical Review of World Energy June 2013* (Houston: BP, 2013), http://www.bp.com/content/dam/bp-country/fr_fr/Documents /Rapportsetpublications/statistical_review_of_world_energy_2013.pdf.

12. Daniel Yergin, "Energy Security in the 1990s," *Foreign Affairs* 77, no. 1 (1988): 111.

13. Samuel J. Locklear III, "The Asia-Pacific 'Patchwork Quilt,'" speech to the Asia Society, Washington, December 6, 2012.

14. Asian Development Bank, "Intense Climate-Related Natural Disasters in Asia and the Pacific," *Learning Lessons*, April 2012, http://www.adb.org/documents/learning -lessons-intense-climate-related-natural-disasters-asia-and-pacific.

15. Partnerships in the Environmental Management of the Seas of East Asia, "Natural Disaster Events in East Asia (1985–2004)," *Tropical Coasts* 12, no. 1 (2005).

16. Bryen Bender, "Chief of US Pacific Forces Calls Climate Biggest Worry," *Boston Globe*, March 9, 2013.

17. Carlyle Thayer, "Chinese Assertiveness in the South China Sea and Southeast Asian Responses," *Journal of Current Southeast Asian Affairs* 30, no. 2 (2011): 77–104.

18. Michael D. Swaine and M. Taylor Fravel, "China's Assertive Behaviour Part Two: The Maritime Periphery," *China Leadership Monitor*, no. 35 (2011): 1–29; Alastair I. Johnston, "How New and Assertive Is China's Assertiveness?" *International Security* 37, no. 4 (2013): 7–48.

19. Hillary Clinton, "Remarks at Press Availability," Hanoi, July 23, 2010.

20. International Crisis Group. "Stirring Up the South China Sea (I)," *Asia Report*, no. 223 (April 23, 2012), http://www.crisisgroup.org/~/media/Files/asia/north-east-asia/223 -stirring-up-the-south-china-sea-i.pdf.

21. White House, *Sustaining US Global Leadership: Priorities for 21st Century Defense* (Washington, DC: White House, 2012); Samuel J. Locklear III, speech delivered at Surface Navy Association Conference, Aiea, Hawaii, January 15, 2014, http://www.andrewerickson .com/2014/01/transcript-of-admiral-locklears-speech-at-the-surface-navy-association -conference/.

22. Barack Obama, "Speech to Parliament," *The Australian*, November 17, 2011.

23. Leon Panetta, address to the Eleventh Shangri-La Security Dialogue," Singapore, June 2, 2012.

24. Dan Blumenthal, Randall Schriver, Mark Stokes, L. C. Russell Hsiao, and Michael Mazza, *Asian Alliances in the 21st Century* (Washington DC: Project 2049 Institute, 2010), http://project2049.net/documents/Asian_Alliances_21st_Century.pdf; Abraham M. Denmark and James Mulvenon, eds., *Contested Commons: The Future of American Power in a Multipolar World* (Washington, DC: Center for a New American Security, 2010), http://www.cnas .org/files/documents/publications/CNAS%20Contested%20Commons_1.pdf.

25. Kevin Rudd, "A Maritime Balkans of the 21st Century? East Asia Is a Tinderbox on Water," *Foreign Policy*, January 31, 2013, http://foreignpolicy.com/2013/01/30/a -maritime-balkans-of-the-21st-century/; Hugh White, "Caught in a Bind That Threatens an Asian War Nobody Wants," *Sydney Morning Herald*, December 26, 2012.

26. James Manicom, *Bridging Troubled Waters: China, Japan, and Maritime Order in the East China Sea* (Washington, DC: Georgetown University Press, 2014).

8 What Threats Do Terrorism and Insurgency Pose in the Asia-Pacific?

Christopher Paul and Nick Nelson

Reader's Guide

This chapter explores the issues of terrorism and insurgency in the Asia-Pacific. It defines "terrorism" and "insurgency" and describes different forms they can take, along with detailing some of the key actors and their motivations, and how those have changed over time. It explores how the region's states have responded to terrorist and insurgent threats, relates terrorism and insurgency to the different security theories introduced in the introduction, and provides a discussion of the factors that can increase the risk of terrorism or insurgency with an eye toward future threats in the region.

Introduction

The purpose of this chapter is to introduce the concepts of terrorism and insurgency in the Asia-Pacific. The chapter begins by providing definitions of the contested concepts of terrorism and insurgency. It then outlines the development of these concepts within the region after World War II, during the Cold War, and after the September 11, 2001, terrorist attacks on the United States. Consideration is then given to the ways in which the threat of terrorism and insurgency varies across the region, how different states respond to these threats, and the factors that increase the risk of terrorism or insurgency.

What Are Terrorism and Insurgency, and What Threats Do Terrorists and Insurgents Pose?

Terrorism is not a new phenomenon. History is replete with examples of leaders being assassinated and cities being destroyed in the name of a "greater good." Despite this, a broadly agreed-on definition of terrorism remains elusive. Even the United Nations has been unable to come up with a definition that is acceptable to the international community. About the only universally agreed-on attribute of the term "terrorism" is that it is pejorative—even terrorists do not like to be called terrorists![1] As a consequence, terrorism is a contested concept, and there are hundreds of "official" definitions that emphasize different attributes of the concept.

While a perfect definition may be elusive, for both theoretical and practical purposes, we must at least try to reach a "good enough" one. For this chapter we bring together some of the most commonly agreed-on components of terrorism to define it as "the deliberate creation and exploitation of fear through violence or the threat of violence in the pursuit of political change."[2]

In seeking its goals, terrorism can be categorized in a variety of ways. One motivation-based typology categorizes terrorism in four categories: **state terrorism**, committed by governments, either internationally or domestically, against perceived enemies; **dissident terrorism**, committed by nonstate groups against governments and other perceived enemies; **religious terrorism**, committed by groups for the greater glory of the faith; and **criminal terrorism**, committed by groups motivated by sheer profit or a combination of profit and political motives.[3] Dissident terrorism is the traditional form, most commonly committed as part of a broader insurgency. Positivist theories (like realism and liberalism) are less likely to be concerned about the motivation of terrorism, viewing terrorism as part of the anarchy that threatens all insufficiently strong international systems. In post-positivist approaches such as constructivism, the motive of the terrorists matters a great deal. The frequently quoted statement that "one person's terrorist is another person's freedom fighter" is a good example of how a security threat can be interpreted differently depending upon who is constructing it.[4]

Like terrorism, **insurgency** is not a new phenomenon and has probably been the most prevalent form of armed conflict since the establishment of organized political communities. This prevalence holds true in the contemporary era, with more than twenty insurgencies ongoing in countries such as Afghanistan, Iraq, Pakistan, the Philippines, and Thailand.[5]

Like terrorism, insurgency often conjures up widely different interpretations, and thus a consensus definition remains elusive. For both theoretical and practical purposes, here we provide a "good-enough" definition of insurgency as "a protracted violent conflict in which one or more groups seek to overthrow or fundamentally change the political or social order in a state or region through the use of sustained violence, subversion, social disruption, and political action."[6] This definition is useful regardless of which broader theory of international relations you choose to employ. Under realism or liberalism, the relevant "political or social order" will be the state itself, and insurgency is either the product of weaknesses within the state or something encouraged by a competitive neighboring state to undermine the first state. Under the constructivist or critical approaches, the political or social order under attack may be more complicated, and may involve the system of ethnic or gender inequality, the system of political representation, or concerns about the distribution of wealth.

With this definition, it should be noted that in seeking to radically change the existing order, insurgents use a number of means, violence being only one of them. In addition to violence, an insurgency also involves political, economic, and social means that are not designed to defeat armed forces but to address the heart of the issue. In this sense, an insurgency is a strategy that comprises a number of means, both violent and nonviolent, to slowly erode the authority and legitimacy of the ruling power.[7]

One of these approaches is terrorism, which falls very much under the violent component of an insurgency. When thinking about the relationship between insurgency and terrorism, you can consider terrorism to be one of many tactics that an insurgency can use in pursuit of its goals. By way of example, the 2004 bombings of the Madrid train stations were a terrorist tactic used by Al Qaeda elements to influence the vote for a new Spanish government that would withdraw troops from Iraq. This was part of a broader insurgent strategy to achieve the expulsion of foreign forces from Islamic societies.

How Did Terrorism and Insurgency Develop in the Asia-Pacific?

While the history of terrorism and insurgency is at least as old as the history of governments and armed conflicts, the modern history of both in the Asia-Pacific can be reasonably said to have begun in the aftermath of World War II. In the postwar period, there were two primary motives for terrorism and insurgency: **anticolonialism** and international **communism**.

World War II strained European colonial bonds in the Asia-Pacific. Many colonies were neglected as their European masters focused their attention on the war in Europe. Japan captured or occupied several colonies, demonstrating that the bond between colony and colonial power was not unbreakable. In many of those occupied colonies, native forces (often encouraged by colonial remnants) resisted the Japanese, gaining both a taste for independence and experience of resistance. When the war ended, many colonial powers bowed to popular pressure for independence in the colonies. In other colonies, European powers reasserted their authority, but faced insurgencies seeking independence born from resistance movements that they themselves had encouraged during the war. Terrorist and **guerrilla** tactics (troops dispersed into small groups to strike at enemy weak points, withdrawing before larger formations of regular troops could be brought to bear) were common. In the immediate postwar period, several colonies demanded and were granted independence: Burma, in 1948; Ceylon (now Sri Lanka), in 1948; and the Philippines, in 1946. Others colonies fought the returning Europeans and won, including Indonesia (1949), and French Indochina, which was partitioned so that part became the independent nations of Cambodia and Laos while the remainder was incorporated into North and South Vietnam (1954). A realist perspective would likely describe these colonial insurgencies as an example of anarchy in action; World War II weakened the European powers to such an extent that they were unable to protect their distant holdings from chaos. Liberal or constructivist approaches would instead ascribe some importance to the specific motives of the insurgents, nationalist, anticolonial, or otherwise.

The other major motive for insurgencies in the Asia-Pacific in the postwar period was international communism. Inspired by the success of first the Russian communist revolution (1917) and then the Chinese communist revolution (1949), regional communist movements were sometimes mixed with anticolonial struggles inspired by **nationalism.** This meant that sometimes they acted in opposition to already-independent governments. Many communist insurgent groups received military, economic, and political support from communist parties or governments in other countries. Such support allowed communist insurgencies to field military forces beyond irregulars and guerrillas. These forces were equipped with heavy weapons and military vehicles and were able to mass in conventional formations. While a wave of communist insurgencies began shortly after World War II, communism continues as a motive for insurgency in the region. Insurgencies involving a significant communist element include the conflict in Indochina (1946–55), the Philippine Huk Rebellion (1946–56), and the insurgencies in Malaya (1948–55), Laos (1959–75), Vietnam (1960–75), Cambodia (1967–75), Kampuchea (1979–92), and Nepal (1996–2006).

The communist and nationalist insurgents in Indochina (what the French called the colonial region that is now Laos, Cambodia, and Vietnam) fought French forces, the forces of the various fledgling national governments, and the forces of the United States, among others, between 1946 and 1975. These insurgents, called the Viet Minh, endured numerous setbacks early in their conflict with the French. They discovered the hard way that French forces would prevail in any relatively evenly matched, pitched battle, and consequently they embraced the art of guerrilla warfare, learning to conduct hit-and-run raids and ambushes of French convoys on the long, narrow roads through the jungle. These communist insurgents made a virtual fetish of camouflage and concealment, protecting themselves from observation and attack by air forces and approaching ground forces virtually unseen. As Fall observed, "Every regular Viet-Minh soldier on the march carried a large wire-mesh disk on his back and head, adorned with the foliage of the terrain through which he was passing. As soon as the terrain changed, it was the responsibility of each soldier to change the camouflage of the man ahead of him as the surroundings changed."[8]

The Viet Minh also dominated government and Western forces in terms of intelligence. Because government forces were often required to move en masse in heavy vehicles and through thick jungle, and because the government and foreign camps were heavily penetrated by insurgent sympathizers, the Viet Minh always knew when these forces left their camps and bases and were approaching. Government forces, conversely, were constantly surprised by the Viet Minh.

Though they were highly effective guerrillas, the secret to the success of these various insurgencies was not guerrilla forces alone. The flow of supplies from North Vietnamese and Chinese communists to these guerrilla forces (down the famous Ho Chi Minh Trail along the border between Laos, Cambodia, and Vietnam) was essential to their success, as it allowed them to eat and replenish ammunition and otherwise receive the supplies they needed in order to continue fighting. Of greatest importance, however, was the participation of regular North Vietnamese forces in the conflict. In the conflicts in Vietnam, Laos, and Cambodia, the presence of both regular North Vietnamese units and well-provisioned indigenous insurgents provided a particularly thorny dilemma for government forces and their supporters. To fight the insurgents, government forces wanted to spread out and to operate in smaller units in order to protect isolated population centers and counter the small unit tactics of the enemy. This was impossible, however, because such a distribution of forces would result in large North Vietnamese formations crushing these isolated small units, referred to in military parlance as "being defeated in detail." In order to avoid conventional defeat, government forces usually massed to oppose the regular units, allowing the insurgent guerrillas to operate with a free hand.

The Cold War was a period of sustained tension between the Western powers (members of the North Atlantic Treaty Organization and other allies) and the Eastern powers (the USSR and the members of the Warsaw Pact) that ran from the late 1940s through to 1989. Although the bulk of the participants and the greatest threat of mutually assured nuclear annihilation were focused in Europe, conflicts associated with the Cold War played out all over the globe, including in the Asia-Pacific.

During the Cold War communism grew in importance as a motive for insurgency as communist countries made the export of revolution a priority. The Western powers usually sought to oppose this spread, due to general opposition to communism and to counter the ambitions of their Cold War rivals. Thus contests over the control of the state, which in other eras of history would have been left primarily to the citizens of that state, took on an international dimension. Examples include the related conflicts in South Vietnam (1960–75), Laos (1954–75), and Cambodia (1970–75). In this era, realism dominated as the primary theory of international relations. Under this view, international communism was seen as little more than an excuse for communist states to undermine and invade their noncommunist neighbors, and any possible legitimate grievances a people might have against their state that might be considered under other theories was overlooked.

In this era, anticolonialism remained a motive for insurgency and was joined by broader nationalist currents that led regionally, ethnically, or religiously distinct groups to seek independence, not just from distant powers, but also from powers closer to home, hoping to carve smaller separate nations off from existing ones. Examples of insurgencies inspired by nationalism and with the goal of **separatism** include the insurgency that led to the formation of Bangladesh in 1971; the movement in Aceh, Indonesia, that began in 1976; the resistance to the annexation of Timor-Leste by Indonesia from 1975 to 1999; and the Tamil insurgency in Sri Lanka that started in 1976.

During the Cold War, the use of terrorism outside the context of an insurgency also became more prevalent. While there were previously attacks that met the definition of terrorism, they were mostly part of the wider range of activities that an insurgent group would undertake in its campaign against the state. As a new development, small groups representing political interests (their own or those of a state sponsor) would stage terrorist attacks (bombings, hijackings, attacks with guns and knives) to call attention to their issue or as demonstrations of capability.

The Cold War ended in 1989 with the historic and symbolic destruction of the Berlin Wall. This substantially diminished communism as a motive for insurgency and the support a communist insurgency could expect from regional communist countries. However, existing communist movements continued their struggles after the Cold War, most notably in Nepal, where Maoist communist insurgents instigated an insurgency in 1996. These insurgents opportunistically joined with antiregime (but prodemocracy) forces when hostilities were resolved in 2006 to gain significant representation in the new government.

Terrorists became more dangerous after the Cold War. Although the number of governments actively supporting terrorists declined, the range of weapons available had increased to the point that terrorists were able to arm themselves effectively at modest cost. Improved armaments, coupled with improved information technology (global positioning systems, satellite imagery, secure communication) made terrorists more lethal. While, previously, terrorist groups may have been constrained by the wishes of their sponsors (who, if displeased, might remove sponsorship), independent terrorist organizations were now largely unconstrained, and the Asia-Pacific saw an increase in the number of mass-casualty terrorist attacks.[9] One such attack was the August 1998 bombing at the main railway station in Guwahati, India, which was carried out by the separatist United Liberation Front of Assam, killing two and injuring twenty.

The world changed on September 11, 2001. The coordinated terrorist attacks that shattered the New York and Washington skylines and the rural calm of Pennsylvania were the most deadly in history (almost 3,000 were killed). The attacks on 9/11 also fundamentally changed how the world thought about terrorism, as perceptions of the potential scope and severity of terrorism increased considerably. And 9/11 crystallized the perception of terrorism as a **transnational** challenge. While Al Qaeda affiliates (e.g., Jemaah Islamiyah) and other transnational terrorist organizations had already been supporting, encouraging, and conducting terrorist activities in the Asia-Pacific, 9/11 brought that fact into popular awareness. Connections between Islamic terrorist organizations suddenly appeared much more sinister, and this increased the salience of the threat from terrorism. Governments threatened by insurgencies sought to find connections between those insurgents and international terrorist organizations so they could claim international support and avoid condemnation for repressing dissent. Coupled with resources made available as part of the United States–led "Global War on Terror," numerous Asia-Pacific governments were able to secure international resources and support for law enforcement and military efforts against terrorist organizations.

Trends in motives for terrorism and insurgency in the region after 9/11 have followed the patterns begun after the Cold War (table 8.1). Communism as a motive for terrorism has continued to decline, and nationalist, **ethnic**, and other types of identity-based separatism have increased. Prominent in the Asia-Pacific has been the mobilization of religious identities for separatist claims, particularly Islamic identities and desires to self-govern under Sharia (Islamic religious) law. The ability of positivist theories of international relations to explain such events has declined, given that terrorists and insurgents are less likely to be acting as proxies for other states, making their characterization as just part of broader state-to-state competition less plausible. This, in turn, has led to the rise of other theoretical perspectives.

How Does the Terrorist Threat Vary across the Asia-Pacific?

The Asia-Pacific's ethnic and religious **pluralism**, along with the problems associated with **decolonization**, have ensured that, at least since World War II, terrorism and insurgency have been a very real part of the landscape. Since 9/11, the uneven spread of **modernization** and the rise of Islamic extremism have ensured that the threat remains, albeit under a different guise. How this manifests itself varies across the region.

In many parts of South Asia, insurgency and terrorism have become a way of life. While, traditionally, such conflicts were motivated by nationalism or separatism, the most serious contemporary challenge to security in this subregion comes from the threat of Islamist militants. This threat evolved as a consequence of Pakistan's Inter-Services Intelligence (ISI) Agency's role, following a realist perspective, as a conduit for billions of dollars of US and Saudi funding to arm and train Islamist militants, including the Taliban, fighting against the USSR during the war in Afghanistan in the 1980s.[10]

After that war, Pakistani society was radicalized and a **jihadist** culture created by the more than 2 million displaced Afghans in Pakistan, along with the numerous terrorist groups that the ISI had created.[11] This jihadist culture created

Table 8.1 Recent Terrorist Attacks in the Asia-Pacific

State	Date	Type	Fatalities
Indonesia	May 28, 2005	Religious, dissident	22+[a]
Indonesia	October 1, 2005	Religious	20+[b]
India	July 11, 2006	Religious, dissident	209[c]
Pakistan	October 18, 2007	Dissident	139+[d]
Pakistan	December 27, 2007	Dissident	24[e]
India	May 13, 2008	Religious	63[f]
China	August 4, 2008	Dissident	16[g]
India	October 30, 2008	Dissident	81[h]
India	November 26, 2008	Religious, dissident	165+[i]
Philippines	November 23, 2009	Criminal, state-supported	58[j]
Philippines	April 12, 2010	Dissident	15[k]
India	May 28, 2010	Dissident (alleged)	145[l]
China	July 30–31, 2011	Dissident	19+[m]
Philippines	January 24, 2012	Dissident	15[n]
Thailand	March 31, 2012	Dissident	16[o]
Pakistan	January 10, 2013	Religious	100+[p]
China	May 22, 2014	Dissident	31+[q]
Pakistan	December 16, 2014	Religious, dissident	145[r]
Thailand	August 17, 2015	Unknown, likely dissident	20[s]

Note: This list includes all the attacks since 2005 that killed fifteen or more people. If more than four attacks took place in a country in that period, only those with the highest casualty figures are featured here.

[a]Associated Press, "Bomb Blasts Kill at Least 22 in Indonesia," NBC News, Palu, 2005, http://nbcnews.com/id/8010669/.

[b]John Aglionby, "Suicide Attacks Leave 22 Dead and 130 Injured," *The Guardian* (Kuta), October 3, 2005), http://theguardian.com/world/2005/oct/03/indonesia.alqaida.

[c]"India Police: Pakistan Spy Agency behind Mumbai Bombings," CNN, New Delhi, 2006, http://edition.cnn.com/2006/WORLD/asiapcf/09/30/india.bombs/index.hrtml.

d Nagesh Narayana and David Cutler, "Chronology: Attacks in Pakistan since July 2007," Reuters, New York, 2007, http://reuters.com/article/2007/12/27/idUSSP260961.

e "Bhutto Exhumation OK, Pakistan Official Says," CNN, Islamabad, 2007, http://edition.cnn.com/2007/WORLD/asiapcf/12/29/bhutto.death/index.html.

f Randeep Ramesh, "Indian Mujahideen Claims Responsibility for Jaipur Blasts," The Guardian (Delhi), May 15, 2008, http://theguardian.com/world/2008/may/15/india.

g Richard Spencer, "Beijing Olympics Security Fears as Terror Attack Kills 16 Chinese Policemen," The Telegraph (Beijing), 2008, http://telegraph.co.uk/sport/olympics/2496900/Beijing-Olympics-security-fears-as-terror-attack-kills-16-Chinese-policemen.html.

h PTI, "Assam Blast Toll Rises to 81," Outlook India (Guwahati), 2008, http://outlookindia.com/news/item.aspx?627571.

i "Mumbai Attacks: One Year On," BBC News (Mumbai), 2009, http://news.bbc.co.uk/2/hi/south_asia/8379828.stm.

j Human Rights Watch, "They Own the People": The Ampatuans, State-Backed Militias, and Killings in the Southern Philippines (New York: Human Rights Watch, 2010), http://hrw.org/sites/default/files/reports/philippines1110.pdf.

k "Philippine Island in Lockdown after Siege," News (Sydney: News.com.au, 2010), http://news.com.au/breaking-news/philippine-island-in-lockdown-after-siege/story-e6frfku0-1225853641789.

l "India Train Attack Death Toll Rises," Al Jazeera (Doha), 2010, http://www.aljazeera.com/news/asia/2010/05/2010529411450630.html.

m Barbara Demick, "Uighur Violence in China Leaves at Least 19 Dead," Los Angeles Times, August 1, 2011, http://articles.latimes.com/2011/aug/01/world/la-fg-china-kashgar-20110801.

n CNN Wire Staff, "Gunmen Attack Fishing Boats in Philippines, Killing 15," CNN, London, January 24, 2012, http://edition.cnn.com/2012/01/24/world/asia/philippines-fishermen-shooting/.

o International Crisis Group, Thailand: The Evolving Conflict in the South (Bangkok: International Crisis Group, 2012), http://crisisgroup.org/~/media/Files/asia/south-east-asia/thailand/241-thailand-the-evolving-conflict-in-the-south.pdf.

p "Quetta Blast Toll Hits 100," The News (Quetta), 2013, http://thenews.com.pk/article-83302-Quetta:-Alamdar-Road-blasts-death-toll-rises-to-84.

q Associated Press, "Urumqi Car and Bomb Attack Kills Dozens," The Guardian (Urumqi), May 22, 2014, http://www.theguardian.com/world/2014/may/22/china-urumqi-car-bomb-attack-xinjiang.

r Declan Walsh, "Taliban Besiege Pakistan School, Leaving 145 Dead," New York Times, December 17, 2014, http://nytimes.com/2014/12/17/world/asia/taliban-attack-pakistani-school.html.

s "Bomb Toll Revised: 20 Dead, 125 Injured," Bangkok Post, 2015, http://bangkokpost.com/news/security/659848/bomb-toll-revised-20-dead-125-injured.

a plethora of radical Islamic terrorist groups that were funded by Afghanistan's opium production and made Pakistan a major exporter of terrorism.[12]

Demonstrating the somewhat effective application of a realist approach, the ISI initially maintained some control over these groups as the provider of funding and equipment and purportedly used them for its own purposes (continued application of realism, and somewhat effective). However, since 9/11 a lack of funding from the Pakistani government, along with diverging goals and interests, has diminished this control. Under realism, a state actor that stops supporting nonstate proxy actors might expect them to wither away. In reality, the jihadist groups have begun to pursue their own policies and goals, and have often acted against the interests of their creators by killing prominent figures, seizing large areas of land in Northern Pakistan, and launching attacks on India.[13] A constructivist or other post-positivist security theory provides a better understanding of why and how these groups escape the control of their state sponsors. In 2001 the Lashkar-e-Taiba and the Jaish-e-Mohammed terrorist organizations attacked the Indian Parliament, and in 2007 the Lashkar-e-Jhangvi, an Al Qaeda–linked terrorist group, claimed responsibility for the assassination of Pakistani prime ministerial candidate Benazir Bhutto. In 2008 a group of ten Lashkar-e-Taiba terrorists conducted a spectacular attack against a number of targets in Mumbai. More recently, in December 2014, the Tehreek-e-Taliban Pakistan claimed responsibility for an attack on a military-run school in Peshawar, Pakistan, which killed more than 140 people, most of whom were children. Given this, it is reasonable to suggest that South Asia is one of the most dangerous subregions in the world today, particularly Pakistan, which has seen the expansion of terrorist organizations, nuclear proliferation, drug smuggling, and an unstable government.

In contrast to South Asia, though the threat of terrorism in Northeast Asia exists, it is relatively minor. North Korea has a demonstrated history as a state sponsor of terrorism. Resentment of the United States has motivated North Korea to provide training, weapons, and technology to nonstate actors, such as Hezbollah, which engage in terrorist activities against the United States and its allies in the Middle East. The dire economic situation in North Korea also means that supporting terrorism, particularly against the United States, provides much-needed currency for Pyongyang.[14]

Despite being one of the largest and most populous countries in the world, the People's Republic of China (hereafter, China) remains relatively free from the domestic threat of terrorism and insurgency, with the exception of the **ethnonationalist** insurgency based in the Xinjiang region of western China. As part of a broader strategy to resist Chinese control and to push for greater **autonomy** of the region, the local population, predominantly the Uyghurs, have undertaken a number of terrorist attacks across China, predominantly in the Xinjiang Uyghur Autonomous Region of Northwest China but also in Beijing.[15] While this is likely to continue in the near future, it does not pose a substantial threat to the security of China or, indeed, the broader region.

Following widespread decolonization in Southeast Asia after 1945, a variety of internal ethnonationalist and religious militant groups emerged, predominantly in response to the failure of governments within the subregion to recognize the rights of minority groups. As part of a broader insurgency, many of these groups used terrorist tactics in pursuit of their aims. While a number of the groups have either been defeated (Malaysia) or made little headway (Thailand), others, such as those involved in the Moro rebellion in the southern Philippines and the Aceh rebellion in Indonesia, continue to challenge central governments.

Since the 1990s, and more recently following 9/11, the threat of terrorism in Southeast Asia has both increased and changed, and many have labeled it the "second front" in the global "War on Terrorism."[16] This is predominantly a consequence of two key factors: first, the vigorous pursuit and unequal implementation of modernization by nation-states within the subregion has resulted in poverty, inflation, and unemployment, which has produced strong feelings of dissatisfaction toward central governments; and second, significantly increased cross-border interaction has allowed criminal and terrorist organizations, most notably Al Qaeda, to penetrate what is the world's largest Muslim region, giving rise to Islamic extremism.[17] The strong links that Al Qaeda has developed with regional Islamist militant groups such as the Abu Sayyaf group and the Moro Islamic Liberation Front in the Philippines, and Jemaah Islamiyah in Indonesia, have not only enhanced the operational effectiveness of these regional groups but have also created a global **jihadi** ideology that aims to overthrow national governments and establish an Islamic state within the region.[18]

How Do Asia-Pacific States Respond to Insurgencies?

There is no single way in which Asia-Pacific states have responded to insurgencies. However, it is possible to identify some simple patterns in state responses based on the period of history, national proclivities, and the nature of the threat faced.

Although it may not have been clear at the time, colonial powers opposing anticolonial insurgencies in the Asia-Pacific were fighting against the tide of history. Colonialism in the region had run its course, and earnest nationalistic efforts were bound to succeed eventually.[19] And though the approach to **counterinsurgency** adopted by the colonial powers no doubt contributed to the duration of these conflicts, the outcomes were much more influenced by when and how the colonial power gave up the colony, and to whom. For example, in Indochina (1946–55) the French failed to offer meaningful autonomy to a postcolonial government until they were forced to, losing North Vietnam to the communists and setting the stage for the broader war in Vietnam. The British in Malaya (1948–60), conversely, had already begun to cede autonomy to the inhabitants when the insurgency erupted. The British remained in charge of the counterinsurgency effort until the conflict was resolved. Then they left an autonomous government that was friendly to them.

During the Cold War, various communist countries in the Asia-Pacific supported communist insurgencies in neighboring countries. Adhering to fear of the "domino theory" of communist expansion (which suggested that if one state were allowed to fall to communism, more and more would follow), the United States sought to prop up threatened governments. The most notable example of such efforts was the conflict in Vietnam, with the United States providing 500,000 troops and billions of dollars in economic and defense assistance. Over the course of the conflict, United States–led counterinsurgency efforts included defoliation (clearing forested areas), massive combat sweeps, interdiction bombing (to keep enemy forces from using roads or bridges), and "pacification" (to remove the influence of insurgents from towns and villages). What the campaign never did was to stop the flow of troops and war material that continuously infiltrated from North Vietnam into South Vietnam; nor did it deal with the threat of invasion by North Vietnamese regulars and the constant low-intensity conflict fought by irregular forces in the South.

Another fairly common response to insurgency by Asia-Pacific states is to seek to smash the insurgency with overwhelming force, an approach characterized elsewhere as the "iron fist" approach.[20] This approach emphasizes military action against the active insurgents, seeking to kill or capture them, and largely ignores whatever grievance inspired the insurgency or motivated popular support. This approach has proven successful when sufficient force has been applied. The Chinese quelled the insurgency in Tibet (1959–74) in this fashion, and the victory of Sri Lanka over the Liberation Tigers of Tamil Eelam (1996–2009) predominantly followed this path. However, iron fist efforts have failed elsewhere, such as in Kampuchea (1978–92), East Timor (1975–99), and Nepal (1996–2006). The iron fist approach is consistent with a realist view of an international system defined by anarchy, with sovereign states engaged in a power struggle in order to maximize their interests. As a consequence, the realist approach focuses on the use of hard power, particularly military power, to defeat threats, including those from terrorism and insurgency.

Other insurgencies in the Asia-Pacific have been met with different approaches. Some of the most successful have combined effective military action with reforms and other efforts to address the grievances that sustain the insurgency. One of the keys has been the discriminating application of force: finding ways to attack the insurgents without alienating civilian populations more than necessary. Examples include the region's efforts against the Huk Rebellion in the Philippines (1946–55), Indonesia's campaign against the Darul Islam insurgents (1949–62), and the efforts against the Moro National Liberation Front in the Philippines (1971–96). Such efforts correspond more closely to the liberal way of thinking, which, unlike realism, allows nonstate actors to have interests and to use the state as an arena to compete for these interests.[21] Liberals thus assert that the best way to combat terrorism and insurgency is through a coordinated multilateral approach that utilizes a combination of both hard and soft power.

India has demonstrated a somewhat distinctive approach to counterinsurgency. The Indian approach relies on the massive passive presence of state security forces. When faced with an insurgency, India has habitually stationed very large numbers of troops in the threatened region. Insurgents operate at their own peril; Indian troops are reasonably likely to be present or nearby and to respond. The approach has been described as an effort to suffocate the insurgents through "the saturation of forces."[22] It is difficult to say how effective this approach has been. India's government has not fallen to any insurgencies; nor has India allowed any separatist regions to achieve autonomy. However, India is host to some of the region's longest-running unresolved insurgencies: in its Northeast (1975–ongoing), that of the Naxalites (1980–ongoing), and in Kashmir (1989–ongoing) (table 8.2).

Recent research by the RAND Corporation identified the core elements of successful counterinsurgency since the World War II.[23] RAND found three things that all successful campaigns against insurgents had in common. First, government forces were able to force the insurgents to fight as guerrillas, because an insurgency that is able to field forces that can take on government forces on an equal footing is a particularly tough challenge. Second, they were able to reduce the support insurgents received from a native population or from the government of another country. Third, counterinsurgent forces were flexible and adaptable, changing their tactics and operations based on changes in insurgent tactics or changes in the conditions in the region of conflict. Future efforts to defeat insurgencies in the Asia-Pacific should keep all three of these lessons in mind.

How Do Asia-Pacific States Respond to Terrorism?

The rise of modern mass-casualty terrorism means that developing and implementing an effective strategy that both prevents attacks and eliminates terrorist organizations is a high priority. Over time, and particularly since the events of 9/11, two main approaches have evolved to countering terrorism—the hard approach and the soft approach—mirroring the "iron fist" and motive-focused approaches used in countering insurgencies.

The hard approach involves a state's security apparatus collaborating in surveillance and intelligence activities and then using security forces to directly target terrorists.[24] This targeting could take the form of: unmanned drone strikes; raids by police or military special forces; the denial of safe havens; or cutting

Table 8.2 Insurgencies in the Asia-Pacific

Location	Dates	Type
Indochina	1946–55	Communist
Philippine Huk Rebellion	1946–56	Communist
Malaya	1948–55	Anticolonial
Indonesia (Darul Islam)	1949–62	Religious
Laos	1959–75	Communist
Vietnam	1960–75	Communist
Cambodia	1967–75	Communist
Bangladesh	1971	Separatist
The Philippines (Moro National Liberation Front)	1971–96	Separatist
Timor-Leste	1975–99	Separatist
India (Northeast)	1975–ongoing	Separatist
Aceh, Indonesia	1976–2005	Separatist
Sri Lanka	1976–2009	Separatist
Kampuchea	1979–92	Communist
India (Naxalites)	1980–ongoing	Communist
India (Kashmir)	1989–ongoing	Separatist
Nepal	1996–2006	Communist

sources of funding. By undertaking these types of actions, terrorist activities and groups are disrupted, and terrorist suspects are detained or arrested—and, where appropriate, are subsequently dealt with through the judicial system. Supporters of this approach believe that it ensures that the cost of engaging in terrorist activities, either by state or nonstate actors, outweighs the benefits.[25] They also argue that, in the contemporary environment where terrorist attacks can be catastrophic, a hard power approach must take primacy.

Conversely, the soft approach aims to counter the extremist ideologies that lead to acts of terror. Against states, widely supported diplomatic, economic, and political sanctions are deemed to be the most useful methods of countering state-sponsored terrorism. Against terrorist organizations, the soft approach espouses engagement, education, and negotiation as keys to success.[26] In countering the radical Islamist threat predominant in the Asia-Pacific, the soft approach stresses engagement with the Muslim community and leaders, many of whom are opposed to the terrorist groups, to issue religious decrees, and to develop education and counseling programs designed to provide a broader understanding of terrorists' misrepresentation of Islam.[27] Proponents of the soft approach to terrorism consider that this approach is the most hopeful approach to secure long-term victory over the terrorists.

In countering the growing threat of the Islamist Jemaah Islamiyah (JI) group in Southeast Asia, Singapore, Indonesia, and Malaysia have used a combination of hard and soft approaches collaboratively. Joint surveillance and intelligence sharing, along with a regional strategy to engage with the Muslim community to counter terrorist ideology and the transnational coordination of direct action against terrorist groups, have all proved remarkably successful in both physically reducing JI and reducing its support.[28] Similarly, the combination of legislative change, policing, surveillance, and negotiation has ensured that groups such as the Japanese Red Army and Aum Shinrikyo have ceased to exist in Japan and that remaining threats are very limited in their use of violence.

JI is an Indonesia-based, transnational terrorist network whose claimed goal is to establish an Islamic state encompassing southern Thailand, Malaysia, Singapore, Indonesia, Brunei, and the southern Philippines. JI was officially founded in 1993, but its roots date back to the Indonesian Darul Islam insurgency in the 1940s. JI shares the anti-Western ideology of global jihad espoused by Al Qaeda, and JI's close links with Al Qaeda have made it a willing proxy to attack Western targets.

Estimates of JI's membership numbers vary from five hundred to several thousand, although detailed information on the size and structure of the organization has not been established. To sustain its activities, JI has developed a sophisticated structure of finances and businesses, and it is fully capable of its own fund-raising through membership donations and criminal and business activities. Many JI operatives trained in camps in Afghanistan before 2000, and the organization includes many skilled with explosives, having employed a range of bombs, car bombs, trucks bombs, and suicide bombs in its attacks.

In the weeks after the 9/11 terrorist attacks, the Singaporean authorities uncovered a JI plot to attack Western targets in Singapore. Subsequent investigations highlighted the extent of JI's network, with operational cells being identified in Indonesia, Malaysia, the Philippines, Singapore, and Thailand and also in Australia and Pakistan. The investigations also linked JI to the 9/11 attacks on the United States.

In October 2002 JI was responsible for the bombing of a nightclub district in Bali frequented by Western tourists that killed approximately 200 and injured more than 300. Other bombings believed to have been carried out by JI since 2002 include the bombing of a Marriott Hotel in Jakarta in 2003, the bombing of the Australian Embassy in Jakarta in 2004, and the Bali II bombing of October 2005, in which three suicide bombers exploded bombs killing more than 20 people and wounding more than 100. In July 2009 JI was involved in the bombings of Marriott and Ritz-Carlton hotels in Jakarta that killed 9 and injured more than 50.

In dealing with its ethnonationalist threat, China is likely to maintain its longstanding policy of "noninterference," and any perceived attempt to undermine the stability of China by another nation through the support of terrorists or insurgent groups will not be tolerated. Notwithstanding this, China has recently expanded its engagement in the region and undertaken joint counterterrorist training activities with such countries as Russia, India, Kazakhstan, and Kyrgyzstan. However, counterterrorism cooperation with other nations on this issue, particularly the United States, is likely to remain marginal.[29]

While the use of both the hard and soft approaches in conjunction may, at times, seem contradictory, it acknowledges the fact that terrorism is rarely an either/or proposition and that an effective counterterrorism strategy necessitates the use of a wide variety of options. When combined, the approaches must be balanced so as to be hard enough to disrupt terrorist threats, while at the same time being soft enough to win "hearts and minds."[30] The challenge for nation-states in the Asia-Pacific is in finding the right balance.

Factors That Lead to an Increased Risk of Terrorism or Insurgency in Asia-Pacific States

It is very difficult to predict outbreaks of unrest or instability such as terrorism or insurgency, let alone the timing of terrorist attacks. The things that make a country more likely to experience unrest or to encourage groups to engage in resistance are different from the things that trigger outbreaks. Sean O'Brien of the US Army Center for Army Analysis offers us the "spark and oily rags" metaphor for thinking about the risk of terrorism and insurgency.[31] The different things that make unrest and resistance more likely determine how oily that country's rags of discontent are. But, no matter how oily these rags are, some kind of spark is required for combustion. A country can have very oily rags for a long time, but nothing is quite severe enough to spark an outbreak of unrest. Of course, the magnitude of the spark can matter, too. A country might not have particularly oily rags, but if subjected to a big enough spark, it will still become engulfed in conflict.

Although it is very difficult to predict the timing of unrest, there are conditions that increase a country's risk of terrorism or insurgency (the oil for the metaphorical rags). Research has sought to identify the factors that are correlated with vulnerability to outbreaks of unrest or instability.[32] Several of the factors identified as strong contributors have been prominent in some Asia-Pacific countries, or could become so in the future. The first factor is high unemployment, which increases the risk of unrest, especially when coupled with a **youth bulge** (when a country's average population is very young, with a particularly large number of people between the ages of fifteen and twenty-four years). Young people are the most likely to join or support unrest, especially if they are not in school or are unable to find a job. The second factor is ethnic or religious discrimination or inequality, which increases the risk of unrest by fostering resentment and encouraging those discriminated against to band together for their own protection, which can in turn lead to terrorism or insurgency. The third factor is an existing terrorist or insurgent group, which increases the risk of terrorism or insurgency, as it means there has already been a sufficient grievance in the country to motivate resistance, and a resistance group is already mobilized and organized. The fourth factor is a high availability of weapons, which increases the prospects for unrest, and the risk that such unrest will be violent. Weapons availability is high in countries that have had recent wars or civil wars, in countries that are near countries that have had recent wars or civil wars, or sometimes in countries that are transit countries for arms smuggling or have limited or no legal controls on weapon ownership.

Thus, the Asia-Pacific countries with the greatest risk of future terrorism or insurgency are, first, those that have current unresolved terrorist or insurgent conflicts; second, those that have a disadvantaged ethnic or religious minority; and third, those that have an abundance of young people but lack sufficient opportunities (economic or otherwise) for these young men and women (especially if weapons are abundantly available).

Conclusion

Insurgency is an effort to overthrow or change the political or social order through a variety of means, whereas terrorism is the deliberate creation and exploitation of fear through violence or the threat of violence in the pursuit of political change. Terrorism is a tactic that is often used by insurgents.

Terrorism and insurgency have long been a part of the political landscape in the Asia-Pacific and will remain so in the future. Motives for terrorism and insurgency have changed over time, evolving from communism and anticolonialism in the post–World War II period to a focus on ethnic or religious separatism in the contemporary era. There is no single way in which the region's states respond to terrorism and insurgency, though there are two main approaches: the hard or "iron fist" approach, and the "soft" or motive-focused approach. A balance between the two is usually most effective.

While the decision by individuals or groups to engage in terrorism is a complex and personal process, certain society-level factors increase the risk of terrorism or other forms of unrest. These factors are present in many Asia-Pacific states, so freedom from terrorism and insurgency in the region is unlikely for the time being.

Key Points

- Motives for terrorism and insurgency have changed over time, beginning with anticolonialism and communism after World War II, but transitioning to ethnic and religious separatism more recently.

- The terrorist threat differs across different areas within the Asia-Pacific. The main sources of variation are the degree of state sponsorship of terrorism and the extent of religious or ethnic divisions within specific states.
- Historical research shows that effective counterinsurgency campaigns, whichever broad approach they follow, do three things: first, overmatching the insurgents and forcing them to fight as guerrillas; second, diminishing the support received by the insurgency, either from local populations or from supporting states; and third, remaining flexible and adaptable, adjusting to insurgents' changes in tactics or changes in the context of the conflict.
- Societal risk factors for terrorism or other forms of unrest include high unemployment; a youth bulge; ethnic or religious discrimination or inequality; an existing terrorist or insurgent group; and a high level of weapons availability.

Questions

1. What is terrorism? What is insurgency? What is the difference?

2. What were the two main motives for insurgencies in the Asia-Pacific during the period immediately after World War II? How did these motives change over time, leading up to the present day?

3. What are the key factors that have led to the changing motives for terrorism since 9/11?

4. How does the threat of terrorism differ between South Asia and Northeast Asia?

5. Choose an Asia-Pacific country with which you are familiar. List three factors from this chapter that could contribute to the risk of future insurgency or terrorism, and indicate the extent to which they are present in the country you chose. Based on this list, do you think the country you selected is likely to experience terrorism or insurgency in the future?

Guide to Further Reading

Gottlieb, Stuart. *Debating Terrorism and Counterterrorism: Conflicting Perspectives on Causes, Contexts, and Responses.* Washington, DC: CQ Press, 2010. http://dx.doi.org/10.4135/9781483330822.
    An interesting and thought-provoking book that debates the central issues on the counterterrorism agenda.

Jenkins, Brian M., and John P. Godges, eds. *The Long Shadow of 9/11: America's Response to Terrorism.* Santa Monica, CA: RAND Corporation, 2011.
    This edited volume includes thoughtful discussions of how the world has (and has not) changed since September 11, 2001, and identifies some of the enduring effects of the United States' responses to the terrorist attacks of 9/11.

Paul, Christopher, Colin P. Clarke, Beth Grill, and Molly Dunigan. *Paths to Victory: Lessons from Modern Insurgencies.* Santa Monica, CA: RAND Corporation, 2013.
> A comprehensive study of all insurgencies completed since World War II. Tests various different concepts for fighting insurgencies against the historical record.

Tan, Andrew T. H. *Security Strategies in the Asia-Pacific.* New York: Palgrave Macmillan, 2011. http://dx.doi.org/10.1057/9780230339156.
> This book discusses the post-9/11 complexities of terrorism and insurgency in Southeast Asia and argues that a unique approach is needed.

Notes

1. Louise Richardson, *What Terrorists Want: Understanding the Enemy, Containing the Threat* (New York: Random House, 2006), 19.

2. Bruce Hoffman, *Inside Terrorism* (New York: Columbia University Press, 2006), 40.

3. Gus Martin, *Understanding Terrorism: Challenges, Perspectives, and Issues*, 4th ed. (Thousand Oaks, CA: Sage, 2013), 40.

4. Ibid., 41.

5. Christopher Paul, Colin P. Clarke, Beth Grill, and Molly Dunigan, *Paths to Victory: Lessons from Modern Insurgencies* (Santa Monica, CA: RAND Corporation, 2013), xii.

6. R. Scott Moore, "The Basics of Counterinsurgency," *Small Wars Journal*, May 4, 2007, 2, http://smallwarsjournal.com/blog/small-wars-pleasures.

7. Paul B. Rich and Isabelle Duyvesten, eds., *The Routledge Handbook of Insurgency and Counterinsurgency* (New York: Routledge, 2012), 11.

8. Bernard B. Fall, *Street without Joy: The French Debacle in Indochina* (Mechanicsburg, PA: Stackpole Books, 1964), 65.

9. RAND Database of Worldwide Terrorism Incidents, RAND Corporation, 2009, http://smapp.rand.org/rwtid/search_form.php.

10. Brahma Chellaney, *Asian Juggernaut* (New York: HarperCollins, 2010), 7.

11. Bruce Reidel, "Pakistan and Terror: The Eye of the Storm," *Annals of the American Academy of Political and Social Science* 618 (2008): 34.

12. Chellaney, *Asian Juggernaut*, 7.

13. Sumit Gangulay and Paul Kapur, "The Sorcerer's Apprentice: Islamist Militancy in South Asia," *Washington Quarterly* 33 (2010): 53.

14. Bruce Bechtol, *The Last Days of Kim Jong-Il: The North Korean Threat in a Changing Era* (Dulles, VA: Potomac Books, 2013), 117.

15. Marc Lanteigne, *Chinese Foreign Policy* (Milton Park, UK: Routledge, 2013), 92.

16. Andrew T. H. Tan, *Security Strategies in the Asia-Pacific* (New York: Palgrave Macmillan, 2011), 2.

17. Peter Chalk, "Militant Islamic Extremism in Southeast Asia," in *Terrorism and Violence in Southeast Asia: Transnational Challenges to States and Regional Stability*, ed. Paul J. Smith, (Armonk, NY: M. E. Sharpe, 2005), 25.

18. Peter Chalk, *Rebuilding While Performing: Military Modernisation in the Philippines 2014* (Canberra: Australian Strategic Policy Institute, 2014), https://www.aspi.org.au/publications/rebuilding-while-performing-military-modernisation-in-the-philippines/SR68_Philippines.pdf.

19. Paul et al., *Paths to Victory*.

20. Ibid.

21. Robert Keohane, "The Globalization of Informal Violence, Theories of World Politics, and the Liberalism of Fear," *Dialog-IO*, Spring 2002, 29–43.

22. Sameer Lalwani, "India's Approach to Counterinsurgency and the Naxalite Problem," *Combatting Terrorism Center Sentinel* 4 (2011): 5.

23. Paul et al., *Paths to Victory*.

24. Syed Mohammed Ad'ha Aljunied, "Countering Terrorism in Maritime Southeast Asia: Soft and Hard Power Approaches," *Journal of Asian and African Studies* 47 (2011): 653.

25. Michael Rubin, "Counterterrorism Strategies: Do We Need Bombs over Bridges? Yes—Military Tactics Are Essential for Fighting Terrorism," in *Debating Terrorism and Counterterrorism: Conflicting Perspectives on Causes, Contexts, and Responses*, ed. Stuart Gottlieb (Washington, DC: CQ Press, 2010), 223.

26. Brigitte L. Nacos, "Counterterrorism Strategies: Do We Need Bombs over Bridges? No—Soft Power Trumps Hard Power in Counterterrorism," in *Debating Terrorism and Counterterrorism*, ed. Gottlieb, 218.

27. Melissa Kronfeld, "Fighting Quietly in the Post-Bin Laden Era: The Tools, Tactics & Techniques of Soft Counter-Terrorism in the Twenty-First Century," *Journal on Terrorism and Security Analysis*, Spring 2013, 11.

28. Donald Weatherbee, *International Relations in Southeast Asia: The Struggle for Autonomy*, 2nd ed. (Lanham, MD: Rowman & Littlefield, 2009).

29. US Department of State, *Country Reports on Terrorism 2013*, http://www.state.gov/j/ct/rls/crt/2013/.

30. Gottlieb, *Debating Terrorism and Counterterrorism*, 203.

31. Sean P. O'Brien, "Anticipating the Good, the Bad, and the Ugly: An Early Warning Approach to Conflict and Instability Analysis," *Journal of Conflict Resolution* 46 (2002): 793.

32. Christopher Paul, Russell W. Glenn, Beth Grill, Megan P. Mckernan, B. Raymond, Matt Stafford, and Horacio Trujillo, "Identifying Urban Flashpoints: A Delphi-Derived Model for Scoring Cities' Vulnerability to Large-Scale Unrest," *Studies in Conflict & Terrorism* 31 (2008): 981–1000.

9 How Relevant Are Internal and Nontraditional Security Challenges in the Asia-Pacific?

Alistair D. B. Cook

Reader's Guide

This chapter explores the internal and nontraditional security threats that states and societies face in the Asia-Pacific. It first argues that it is essential to take a more comprehensive approach to security in the region—one that focuses on threats such as separatist movements, pandemics, food riots, and access to water. It then explores the methods that states and societies may use to respond to these security threats. It concludes by evaluating how internal and nontraditional security challenges might pose a broader regional security challenge.

Introduction

The Asia-Pacific is home to different types of government, has a history of colonialism and independence movements, and exhibits varying levels of economic development and democracy. As a result, the region's states face a variety of internal and nontraditional security challenges. To understand these challenges, one needs to look inside states in order to identify the stress factors that generate these challenges and their consequences for states and societies across the region.

To explore the region's internal and nontraditional security challenges, this chapter focuses primarily on Southeast Asia. After World War II, the newly independent states across Southeast Asia were made up of different religions, ethnic groups, and ideologies, which made them potentially unstable. Consequently, Southeast Asia became known as the "region in revolt," with different groups competing for control of states.[1] As a result of this experience, the **Westphalian state system** of clearly defined international borders has faced multiple challenges in the region. Ethnic, religious, and ideological groups straddle state borders and compete for power within them. In response to the fragile nature of Southeast Asian states, many **"soft authoritarian"** governments emerged across the region, which prioritized state stability over the concerns of marginalized groups. The neglect of marginalized groups ultimately laid the groundwork for internal conflicts, some of which remain unresolved today.

Security in Southeast Asia is partly characterized by internal threats to state stability. This characterization differs markedly from other regions, where security has primarily been defined in terms of the ability of states to defend themselves against external aggression, particularly during the Cold War. Alongside internal security challenges, security in Southeast Asia is also characterized by nontraditional security concerns, which are challenges to the survival and well-being of peoples and states. These threats arise primarily from nonmilitary sources, such as climate change, resource scarcity, infectious diseases, natural disasters, irregular migration, food shortages, and transnational crime. These sources of insecurity are often transnational in scope; require comprehensive political, economic, and social responses; and have grown significantly since the rise of **globalization**. Nontraditional security concerns have increased in importance across the region as people have moved in search of jobs and a better life, and trade in goods and services has increased. As this chapter illustrates, Southeast Asia is a major crossroads for trade and the movement of people from all corners of the globe. In a highly interconnected world, security concerns can easily cross state borders. While trade can bring prosperity, it can also facilitate the spread of disease and thus affect the general well-being of populations. We saw this in 2002, with the Severe Acute Respiratory Syndrome (SARS) virus pandemic, which originated in China and spread widely. Within weeks, SARS spread from Hong Kong to infect individuals in thirty-seven countries involving 8,273 reported cases and 775 reported deaths. This means that world leaders must pay close attention to developments in Southeast Asia, for they can easily have a global impact.

In addition to the spread of pandemics, increasing global connectivity means that a poor harvest in one country can lead to food price spikes in another country. In 2008 the Indonesian government was forced to reduce import taxes for soybeans after thousands of people took to the streets to protest rising prices. Other factors that drove up food prices included an increase in the number and severity of floods and other disasters affecting harvests; the use of agricultural land to produce biofuel, which reduces the area of land in which to grow food; larger populations meaning more mouths to feed; and the reduction of exports by food-producing countries due to increased domestic demand. What is clear is that there are significant linkages between different security challenges—from climate change to food security and energy security. This suggests that there is a clear need for a comprehensive approach, which begins by explaining how the security study theories one holds shape how one understands these issues.

Of the two traditional schools of thought—liberalism and realism, broadly defined—realism lends itself to nontraditional security by focusing on the risk of the outbreak of war over access to natural resources and one state's ultimate control over them. Liberalism lends itself to nontraditional security and highlights how states can benefit from cooperating to share access to natural resources and thus avoid the outbreak of war. The traditional schools of thought therefore focus on the geostrategic nature of nontraditional security. Nontraditional schools of thought like constructivism and critical security studies unpack the nature of a security threat, who it affects, and what effects it will have on the functioning of a state and society. These nontraditional schools do not view the state as a static unit but rather as a complex web of interactions between various interests at different levels of analysis—individual, state, regional, and global. As a nontraditional school, human security defines the security referent as an individual rather than the state. As such, it does not emphasize internal or transnational security threats, but instead seeks to protect the individual or community from a threat to their well-being—with well-being defined as freedom from fear of persecution, freedom from want (access to basic needs like food, water, shelter, etc.), and the freedom to live in dignity. When considering internal and nontraditional security, a comprehensive approach allows the researcher to utilize the best tools of both the traditional and nontraditional schools of thought to understand the puzzle at hand.

Internal Security in the Asia-Pacific

Across Southeast Asia, systems of government range from **democracy** in the Philippines and Indonesia, to **communism** in Vietnam and Laos, to an **absolute monarchy** in Brunei, and to a quasi-civilian government in Burma/Myanmar. The way these systems operate, and the dominant ideological or structural composition of the government, determine the government's representativeness and accountability, and its decision-making processes.

How can one best understand how well a particular country functions? This usually includes an analysis of its political institutions (e.g., the strength of the three branches of government—the judiciary, legislature, and executive), its bureaucratic structure (including the nature and effectiveness of the civil service), and its military strength (its armed forces' level of professionalism, effectiveness, and size). A comprehensive security approach also considers levels of societal resilience (e.g., how individuals and communities respond to insecurity) and of economic development (e.g., the standard of living and wealth distribution) to determine a country's strength and stability.

As the states across Southeast Asia became independent after World War II, they largely retained the international borders that were imposed under colonialism. This meant that some historic ethnic communities found themselves on different sides of an international border or were forced to migrate into new areas. For example, the Kachin people can be found in northeastern India, northern Burma/Myanmar, and Yunnan Province in southwestern China. In addition,

Southeast Asian countries were home to large numbers of migrant workers from China and India during the colonial period, and many stayed after independence. This, along with more recent employment-seeking migration, has added a further ethnic dimension to the Southeast Asian states. National independence came in waves across the region: Indonesia gained independence from the Netherlands in 1945; Vietnam from France in 1946; the Philippines from the United States in 1946; Burma from the United Kingdom in 1948; Cambodia and Laos from France in 1953; Malaysia from the United Kingdom in 1957; Singapore from the United Kingdom in 1963; and Brunei from the United Kingdom in 1984. However, there were also periods of regional domination, such as Indonesia's annexation of East Timor from 1975 to 1999, after a brief period of independence from Portugal.

Independent Southeast Asian states are therefore largely **heterogeneous**, which can pose a threat to state stability and the well-being of communities, as different groups compete for power and influence. For example, Indonesia has more than three hundred ethnicities across the archipelago. The Javanese form the largest single group; they constitute approximately 42 percent of Indonesia's population of 237.6 million people. Reflecting its heterogeneity, Indonesia has experienced significant **communal violence**, such as the anti-Chinese violence, which raised global concern about the country's peace and stability.[2] Heterogeneous countries can also be home to **secessionist movements** by pro-independence groups (e.g., the Kachin Independence Army in Burma/Myanmar). These movements dispute the legitimacy of the ruling government and claim the right to break away from the existing state in order to form their own state.

The Southeast Asian states have also faced other internal security challenges, the most notable of which emerged during the Cold War, when communist parties and insurgencies mobilized across the region. Complicating the legacy, some of these groups emerged before states achieved their independence and contributed to ending colonialism. For example, during the Malayan Emergency of 1948–60, the Malayan Communist Party used guerrilla tactics in its attempt to overthrow the colonial authorities and establish a communist state. After the communist revolutions in Vietnam, Laos, and Cambodia, there was a regional fear that the spread of communism would topple newly established states. This fear eventually became one of the dominant motivations for the formation of a regional multilateral organization that could help encourage peace. To achieve this, the Association for Southeast Asian Nations (ASEAN) was established in 1967. It also explains the desire of ASEAN members to uphold the principle of noninterference in another member country's domestic politics and to build a consensus among the member states. This is further discussed in chapter 11.

One of the consequences of the communist threat that still resonates today is the internal security acts in Malaysia (albeit recently reformed) and Singapore. These acts provide for preventive detention, whereby someone suspected of promoting communism (or another ideology deemed threatening to the state) can be arrested and detained for up to one year. The repeal of these acts has become a rallying point for opposition and human rights organizations in Malaysia and Singapore. These groups argue that detention is often arbitrary and contrary to the rule of law, which threatens an individual's personal security.

Across Southeast Asia, the newly independent states also faced ethnic rivalries. There have been attempts to negotiate political settlements between different ethnic and religious groups. For example, though Singapore joined the federation of Malaysia in 1963, it was expelled in 1965 as a result of then-Singaporean prime minister Lee Kuan Yew's rejection of the special privileges granted in the Constitution to Malays over the Chinese and Indian ethnic minorities. These laws would have redistributed economic power to Malays at the expense of the Chinese and Indian ethnic minorities. In extreme cases, ethnic rivalries can lead to large-scale expulsions of a particular ethnic group. This occurred after the 1972 military takeover in Burma/Myanmar that resulted in the widespread expulsion of Indians and saw the subsequent deterioration of India–Burma relations.[3]

In Burma/Myanmar, alongside the souring of relations between the Burmese majority and the Chinese and Indian minority populations, several internal conflicts remain unresolved. Burma/Myanmar is home to numerous ethnic groups, which are known as "national races." The largest group—the Burmese—accounts for approximately two-thirds of the population. The remaining one-third is made up of various ethnic nationalities, many of which have waged war against the majority Burmese since independence.

The members of one ethnic nationality, the Karen, have fought against the Myanmar Army since 1949. This conflict was the world's longest-running civil war, until a shaky cease-fire agreement was signed in 2012. The Karen State lies along the Thailand–Burma/Myanmar border. As a result of this conflict, there are approximately 150,000 displaced people in camps on the border, an estimated 230,000 internally displaced within southeastern Burma/Myanmar, and many more Karen refugees in Thailand. These people face multiple challenges, such as accessing food, water, health care, and education, and also attempting to remain out of harm's way. This illustrates how internal conflict can have significant transnational implications, as many people can be displaced across international borders. Often, governments see the mass movement of people into their country as a destabilizing force, as they fear the conflict might spill over into their state. Governments also worry that local populations could be overwhelmed with a mass influx of foreigners, which could spark an intercommunal conflict.

While ethnicity defines some internal conflicts, religion and access to natural resources can define others, and are often interlinked. Some indigenous communities were displaced and disenfranchised from their ancestral lands as a result of colonization. In some cases, this displacement and disenfranchisement continued after independence and so conflict continues. For example, in Mindanao in the southern Philippines, the Moro Islamic Liberation Front has waged an insurgency, alongside other groups, against the dominant Christian state. This front has sought Moro independence as well as land rights and access to natural resources.[4]

In another example, in southeastern Burma/Myanmar, the Democratic Karen Buddhist Army (DKBA) split from the predominantly Christian-led Karen National Liberation Army in 1994. The DKBA subsequently signed a cease-fire agreement with the Myanmar Army in 1995. As a result of the cease-fire, many members of the DKBA profited from business deals to exploit natural resources in Karen State in exchange for supporting the Myanmar Army.

These cases suggest that it is important to investigate overlapping interests in internal conflicts—such as ethnicity, religion. and access to natural resources—in order to identify the motivations behind a conflict and to understand the underlying internal security threat.

Nontraditional Security Concerns

States in the Asia-Pacific and especially Southeast Asia also experience nontraditional security concerns, which include threats to a particular community or transnational threats that cross state borders and affect more than one state and community. This section investigates environmental, health, food, and water security concerns across the region. Because the security threats in the region frequently overlap, a comprehensive approach to security makes it possible to identify the connections between various nontraditional security threats.

Environmental Security

Across Southeast Asia and the broader Asia-Pacific, the role and impact of climate change on states and societies are becoming ever more apparent. In 2013 the Intergovernmental Panel on Climate Change predicted that sea levels would rise between 26 and 82 centimeters within the next century. The two major causes of global sea-level rise are the thermal expansion of the oceans (water expands as it warms) and the loss of land-based ice due to increased melting.[5] Because many states in the region are either archipelagic or coastal, rising sea levels are an important determinant of vulnerability.[6]

While many states and communities across Southeast Asia are aware of the environmental threats posed, state and societal responses remain varied. What rising sea levels mean for low-lying island states is that coastal communities need to be either partly or wholly relocated. There will also be fundamental changes in island geography, settlement patterns, subsistence systems, societies, and economic development.[7] This can reshape the function and form of low-lying states as they try to adapt to the new environment. The effects of climate change are a major security concern for low-lying island states, because they have an impact not only on the well-being and human security of their populations but also on their basic existence as island nations.

Further afield, in larger developed countries like Australia, New Zealand, and the United States, policymakers are considering the options available if sea-level rise necessitates large movements of people outside their home countries. Where can they resettle? What does this mean for a potential recipient country like the United States? This illustrates the complexity of security, whereby a global phenomenon like climate change can have a significant impact on a local community like a tiny, low-lying island state, which can subsequently have an impact upon a large, remote state like the United States.

The Asia-Pacific is also known as the "ring of fire" because the world's largest number of earthquakes and volcanic eruptions occur in the basin of the Pacific Ocean. When earthquakes, typhoons, and volcanic eruptions occur, they of course have an impact on the communities in their path. When this happens, it is called a natural disaster, which is defined as "a situation or event which overwhelms local capacity, necessitating a request to a national or international level for external assistance; an unforeseen and often sudden event that causes great damage, destruction, and human suffering" (table 9.1).[8]

Costs related to natural disasters worldwide have increased from $50 billion a year in the 1980s to $200 billion in the last decade. Yet in 2014 only 4 percent of spending for natural disasters was used for prevention and preparedness, with 96 percent spent on response.[9] While disaster-affected communities could move to safer areas or adapt where they are to such events, in the vast majority of times, communities and governments fail to act, which exacerbates the impact of the natural disaster. Instead, communities and governments respond only after the disaster happens.

For example, in November 2013 Typhoon Haiyan hit the Philippines, causing 6,000 deaths, affecting 11 million people, and costing an estimated $10 billion—the largest number of recorded deaths resulting from an extreme climatic event. It also caused the most significant level of destruction of infrastructure on record for the third-most-exposed country worldwide.[10] However, though there was early warning of the typhoon, few anticipated its ferocity. The government also had a limited capacity to respond and relied heavily on international humanitarian and development assistance. The Asian Development Bank estimates that up to 74 percent of the population of the Philippines is vulnerable to disasters, with typhoons causing the largest number of deaths (31,373 in total, and an average of 1,000 deaths per year) and affecting the largest population (9.3 million). The annual cost of disasters to the Philippines economy is between 0.7 percent and 1 percent of gross domestic product. It identifies that storms are the dominant risk, with an annual average loss of $151.3 million—followed by floods, cost-

Table 9.1 Recent Natural Disasters in Asia

State	Date	Type	Fatalities
Southeast Asia	2004	Earthquake, tsunami	227,898+[a]
China	2008	Earthquake	69,197+[b]
Pakistan	2010	Floods	1,752+[c]
Philippines	2013	Typhoon (Haiyan)	6,300+[d]
India, Pakistan	2015	Heat waves	4,500+[e]

[a]Earthquake Hazards Program, *Magnitude 9.1: Off the West Coast of Northern Sumatra* (Washington, DC: US Geological Survey, 2004), http://earthquake.usgs.gov/earthquakes/eqinthenews/2004/us2004slav/#summary.

[b]Sina, *Zhi 21 ri Sichuan dizhen zhi 69,197 ren yunan 18,222 ren shizong* (As of the 21st, there are 69,197 victims of the Sichuan earthquake and 18,222 people are missing) (Beijing: Sina, 2008), http://news.sina.com.cn/c/2008-07-21/170415971186.shtml.

[c]Singapore Red Cross, *Pakistan Floods: The Deluge of Disaster—Facts & Figures as of 8 September 2010* (Singapore: Singapore Red Cross Society, 2010), http://redcross.sg/news/pakistan-floods-the-deluge-of-disaster.

[d]National Disaster Risk Reduction and Management Council, *NDRRMC Update: Updates Re the Effects of Typhoon "Yolanda" (Haiyan)* (Quezon City: Republic of the Philippines, 2014), https://web.archive.org/web/20141006091212/http://www.ndrrmc.gov.ph/attachments/article/1177/Update%20Effects%20TY%20YOLANDA%2017%20April%202014.pdf.

[e]Reuters, "India Heatwave: Death Toll Passes 2,500 as Victim Families Fight for Compensation," *The Telegraph* (London), 2015, http://telegraph.co.uk/news/worldnews/asia/india/11645731/India-heatwave-death-toll-passes-2500-as-victim-families-fight-for-compensation.html; Kamran Haider and Khurrum Anis, "Heat Wave Death Toll Rises to 2,000 in Pakistan's Financial Hub," Bloomberg, New York, June 24, 2015, http://bloomberg.com/news/articles/2015-06-24/heat-wave-death-toll-rises-to-2-000-in-pakistan-s-financial-hub.

ing $68.8 million; earthquakes, $33.2 million; volcanic eruptions, $14.9 million; droughts, $14.7 million; and landslides, $1.5 million.[11] All these disasters are of course extremely costly for a developing economy. The international response to Typhoon Haiyan was led by the United Nations Office for the Coordination of Humanitarian Affairs. Since the 2005 Pakistan earthquake, a cluster approach has been used to coordinate the international humanitarian response to a disaster like Typhoon Haiyan. Each cluster (e.g., including health, food, early recovery, and livelihoods) is led by a particular UN agency, such as the World Food Program, which heads the food security cluster. In coordination with this system, individual countries also responded with humanitarian assistance. The United States sent the USS *George Washington*, which arrived in the Philippines to provide medical services and airlift survivors.[12]

The Typhoon Haiyan disaster demonstrated the need for better infrastructure, preparedness, and adaptation plans in the areas most vulnerable to natural disasters, along with anticipation of the level of insecurity their populations will experience as a result. Parts of the Philippines are acutely vulnerable to natural disasters, and in the wake of a disaster they rely on external assistance from the national government, foreign governments, and international organizations. In December 2014 the Philippines was hit by Typhoon Hagupit, which was less ferocious than Typhoon Haiyan. The Philippines had in place better warnings of storm surges, advanced placement of food stock, and medical teams to minimize casualties, and 1.7 million people were successfully evacuated to 5,200 evacuation centers.[13] Ultimately, though the risks of natural disasters pose a significant security threat to the Philippines, better preparedness can help prevent a similar level of devastation from reoccurring. In turn, this can improve the security and stability of the Asia-Pacific.

Health Security

With more trade and greater movements of people, the Asia-Pacific faces the potential spread of pandemic diseases. The outbreak of a pandemic like SARS poses an acute risk to states and societies, for it can strain public services such as hospitals. It can also affect people's ability to work and travel, with subsequent economic effects on business and trade. In addition, the countries across Southeast Asia and the broader Asia-Pacific face chronic health security challenges like the spread of HIV/AIDS, which can strain the capacity of states and societies to provide for the well-being of their populations.

The 1994 United Nations Development Program's *Report on Human Security* identified the major cause of death in developing countries as infectious and parasitic diseases. This is mostly a result of poor nutrition and an unsafe environment. People in the Asia-Pacific suffer a disproportionate burden of communicable diseases when compared with other regions of the world. Of the 14 million annual deaths in the region, 40 percent are due to communicable diseases, compared with the global average of 28 percent.[14] Because the region is home to developing economies that experience these challenges, these diseases are therefore a significant security threat.

The November 2002 outbreak of SARS that began in Guandong, China, highlighted the importance of good health governance and the fragile nature of health systems. SARS quickly spread to several other Asia-Pacific states, infecting about 8,273 people, and causing 775 reported deaths in thirty-seven countries before it was brought under control. This rapid spread revealed to governments across the Asia-Pacific, and particularly those in Southeast Asia, their populations' vulnerability to pandemics. It also demonstrated that even infectious diseases with a very low incidence but high mortality rates can generate significant economic, political, and diplomatic fallout. The growth of China's gross domestic product appears to have been reduced by 3.1 percent in the second quarter of 2003, when the SARS outbreak was at its peak. Hong Kong is estimated to have lost $3.7 billion, and SARS' economic impact was more than $4 billion on Canada, which

Table 9.2 Recent Pandemics in the Asia-Pacific

Principal State(s)	Date	Type	Asia-Pacific Fatalities
China, Hong Kong, Singapore	2002	SARS	774+[a]
Global	2009–10	H1N1 influenza	3,850+[b]
India	2015–ongoing	H5N1 influenza	2,172+[c]

[a] World Health Organization, *Summary of Probable SARS Cases with Onset of Illness from 1 November 2002 to 31 July 2003* (Geneva: World Health Organization, 2004), http://who.int/csr/sars/country/table2004_04_21/en/.

[b] World Health Organization, *Pandemic (H1N1) 2009: Update 112* (Geneva: World Health Organization, 2010), http://who.int/csr/don/2010_08_06/en/. The death toll was calculated by adding the figures for Southeast Asia and the Western Pacific.

[c] PTI, "Swine Flu Claims 5 More Lives, Toll Rises to 2,172," India TV News, New Delhi, 2015, http://indiatvnews.com/news/india/swine-flu-claims-5-more-lives-toll-rises-to-2-172-49647.html

also resulted in an estimated 28,000 lost jobs there, illustrating the transnational nature of health security.[15] Outbreaks of infectious diseases have been occurring more frequently and in new areas, posing significant threats to states and societies worldwide, but particularly in the Asia-Pacific, where health systems are often inadequate (table 9.2).

The SARS crisis also highlighted the shortcomings of national health systems, both in responding to the outbreak and in their ability to cooperate with the World Health Organization and other affected states. The SARS outbreak demonstrated to policymakers and the medical community that an effective pandemic response requires a coordinated, multisectoral approach. Consequently, the Singapore government established a ministerial task force to coordinate its response, and adopted a "whole-of-government" approach that brought together the relevant ministries to decide policy. Further afield, there is a growing trend at the global level for a more robust global health system that facilitates dialogues between health, security, and foreign policy professionals at the local, national, regional, and international levels.

Most recently, the 2014 outbreak of Ebola in West Africa has led to comparisons with the SARS outbreak a decade ago, as Ebola would have the potential to exert extensive economic and human costs if it were to reach Southeast Asia. However, more interest has been raised about how the region's medical expertise could add value to mitigating Ebola's toll on West Africa.[16] The World Bank estimated that Ebola would cost the West African region $32.6 billion by the end of 2015. In the past decade the world has seen two major health crises with significant security implications, which substantiate calls to strengthen regional cooperation in disease surveillance and pandemic preparedness as a key aspect of regional security.[17]

Food Security

Food security is a key concern in the Asia-Pacific. In 2014 the region had the largest number of hungry people compared with all other regions worldwide. The Asia-Pacific Economic Cooperation forum (APEC) estimates that 552 million people in the region, or 13.5 percent of its total population, are hungry. Furthermore, the APEC region is home to countries that are highly dependent on international markets and that are major food exporters.[18] It is therefore essential that cooperative mechanisms be further enhanced to ensure food security in the region.

The rapid economic progress made across the Asia-Pacific is clear, and though it has raised millions out of poverty, it has also created new challenges. The emergence of Bangkok, Beijing, Jakarta, and Manila as major metropolises illustrates the dramatic urbanization that is taking place across the region. People move from the countryside to urban or peri-urban areas in search of jobs and the chance for a better life. In 2008, for the first time, the region's number of people living in cities overtook those living in the countryside. This is particularly important in this region, given that seventeen of the world's top twenty-five most densely populated cities are in the Asia-Pacific. Alongside the massive movement of people come significant security challenges that threaten the stability of these countries. With a rapidly aging population in the countryside, it is increasingly difficult to ensure that enough agricultural workers are available to harvest the crops needed to feed populous urban areas.

The scale of this shift in Southeast Asia is remarkable; the urban population increased from about 15 percent of the total population in 1950 to almost 42 percent by 2010. It continues apace today (table 9.3). Many presumed that these largely agrarian economies would not adapt and that people would instead rely on traditional food markets. However, the region is also experiencing multiple systematic changes, with increasingly competitive supermarkets, large wholesale and retail companies, more consumer choice, and more consumption of processed food. These shifts can pose other health concerns, like increasing rates of obesity and poor health. And they can also increase marginalization, as many foods can now be sourced through global supply chains, which means that many Southeast Asian small-scale farmers and the urban poor continue to live in poverty.[19]

When one evaluates the food security of a population, it can reveal this security's complex nature. Thus, four indicators can assist in the measurement of a population's food security: availability, access, utilization, and stability. These indicators give rise to several questions: Is there enough food to go around? What sort of access does a population have to food? Can everyone access the available food, or are people constrained by price for basic foods? These questions can inform one's assessment of a population's food income or what state a

Table 9.3 The Ten Most Densely Populated Cities in the Asia-Pacific

State	City	Population Density (km²)	State	City	Population Density (km²)
Japan	Tokyo–Yokohama	37,843,000	Bangladesh	Dhaka	43,500
Indonesia	Jakarta	30,539,000	India	Hyderabad	40,300
Philippines	Metro Manila	24,123,000	India	Mumbai	32,400
India	Delhi	23,998,000	India	Kalyan	30,500
Pakistan	Karachi	23,500,000	India	Vijayawada	30,100
South Korea	Seoul-Incheon	23,480,000	Bangladesh	Chittagong	28,500
China	Shanghai	23,416,000	India	Malegaon	28,000
China	Beijing	21,009,000	Hong Kong SAR[a], China	Hong Kong	26,400
China	Guangzhou-Foshan	20,597,000	Macau SAR, China	Macau	25,300
India	Mumbai	17,712,000	India	Aligarh	24,600

[a]SAR = special administrative region.

Source: Demographia Foundation, *Demographia World Urban Areas: 11th Annual Edition—2015.01 (Built-Up Urban Areas or World Agglomerations)* (Belleville: Demographia Foundation, 2015), http://demographia.com/db-worldua .pdf. Includes whole urban areas. Population data estimates from 2010–13 base years. Density data estimates from 2011 base year.

food market is in. These indicators can also help explain the direction in which a food market is heading. Food utilization illustrates the impact of inadequate food intake and poor health, such as child undernourishment. That is, are people getting the recommended daily allowance of basic food groups? And to address food stability, one must ask whether access to food will improve. Are food prices stable? For instance, one reason for the huge recent variations in the price of rice is the increasing frequency of extreme weather events such as floods, droughts, and earthquakes in the Asia-Pacific.[20] This again underlines how one nontraditional security threat can have an impact on another nontraditional security concern.

Overall, the state of food insecurity across the Asia-Pacific varies from country to country. However, interconnections between indicators can be seen. For example, in many countries improved access to food can also lead to better health. This further illustrates the need to take a comprehensive approach to security in order to tackle threats and challenges. What is clear is that food insecurity cannot be addressed solely at the national level but will require cooperation with other states. And because some states are net food producers and others are net food importers, striking the right balance will be key to increasing food security across the region.

Water Security

Water scarcity is becoming more acute in the Asia-Pacific and is being exacerbated by population growth, urbanization, industrialization, energy demand, and climate change.[21] Competition over access to water threatens both states' and human security. Many countries across the region share water resources. Most notably, the Mekong River starts on the Tibetan Plateau in China and follows a course through Burma/Myanmar, Laos, Thailand, Cambodia, and Vietnam. As a result, if one country builds a dam on the Mekong River within its borders, it can affect several countries downstream. This leads to tensions between the downstream and upstream countries. More than 60 million people depend on the Mekong River and its tributaries for food, water, and livelihoods. China has constructed seven large dams and more are under construction in Tibet, Yunnan, and Qinghai. Laos began preliminary construction of the Xayaburi Dam in Northern Laos in 2012 but suspended work on it because of complaints by Cambodia and Vietnam. These dams have altered the course of the Mekong River, which threatens the livelihoods of many people downstream. An estimated 80 percent of the 60 million people who live in the Mekong River Basin rely directly on it for their food and livelihoods.

The effect of the dams alters the natural river course upstream, which means that more salty water can make its way into the lower Mekong River from the sea. This in turn can affect farming irrigation systems. The dams can also block the flow of sediment, which means less fertile soil is available for farmers downstream, and dams can also affect water levels and the number of fish living downstream.[22] Each year, Mekong fish have an estimated first sale value of up to $7 billion, according to the Mekong River Commission.[23] This does not include the subsequent economic effects of logistics and food preparation, or even those who fish every day and live outside the formal economy.

More than fifty years ago, the United Nations founded the Mekong Committee, which focused on developing large-scale projects along the Mekong River. In 1995 the Mekong Agreement was signed, which established the Mekong River Commission (MRC). The MRC is run by its four member countries: Cambodia, Laos, Thailand, and Vietnam. As the effects of large-scale development became more pronounced, the Mekong Agreement shifted the focus from the development of large-scale projects to the sustainable development and management of natural resources.[24] However, China is noticeably absent from the MRC, which limits its effectiveness in addressing the core concerns of those communities reliant on the Mekong River. Though current cooperative measures are limited, it appears that cooperation can offer a way to address the multiple security concerns generated by the river's water.

While transboundary river systems are a key part to understanding water security in the Asia-Pacific, there are also places with limited freshwater access. Developed Asian economies like Hong Kong and Singapore rely on water supplied by, respectively, mainland China and Malaysia. Singapore's relations with Malaysia are colored by water issues because, as an island city-state, Singapore depends on Malaysia for water on the basis of two agreements that expire in 2061. Since achieving independence, Singapore has built into its public policy the need to increase water use efficiency and water catchment in an effort to become water independent by 2061. Thus, Singapore has implemented its "four national taps" strategy, which includes water from four sources, including local catchment areas. Since 2011 these catchment areas have been increased from half to two-thirds of Singapore's land surface. The remaining three national taps are imported water; reclaimed water, known as NEWater; and desalinated water. Projections suggest that by 2060, Singapore will produce 80 percent of the water it needs from NEWater and desalinated water.[25] This illustrates the high level of attention given to water as a key security challenge, even in advanced, developed countries like Singapore.

Conclusion

The first section of this chapter established that many states across the Asia-Pacific face similar challenges to their internal stability. When looking at state stability, it is important to ask whose security is being achieved and by what means. This allows us to understand how an internal conflict is framed. Above all, it can be seen that internal security is a major concern of states across the Asia-Pacific.

The second section of the chapter considered how particular nontraditional security issues—such as climate change, disasters, pandemics, food, and water—are security threats. Each of these issues illustrated the multifaceted and overlapping dimensions of security. Global climate change can have a significant local impact on communities that are vulnerable to sea-level rise and can cause population displacement, with implications for recipient countries like the United States. Likewise, Typhoon Haiyan demonstrated how a natural disaster can lead to death, displacement, and destruction, which of course causes global concern. The SARS outbreak illustrated how a transnational health emergency can have an impact upon a state's ability to function. These examples also demonstrated the human and economic costs of being ill prepared.

The issues of water and food show the crosscutting and interdependent nature of security threats. Overreliance on food imports can make states vulnerable to global food price shocks, which can cause civil unrest. Equally, disasters can have an impact on food production and in turn on global prices. Similarly, when an upstream country like China dams the Mekong River to build a hydropower plant, it can have a significant impact on farming and fishing in countries downstream, causing food insecurity for their populations and making these countries more reliant on upstream states and imports that are vulnerable to the global market.

Of utmost importance is the need to evaluate internal or nontraditional security concerns in the Asia-Pacific by gaining an understanding of the complex web of interactions among the multiple stakeholders. This underlines the need to avoid artificially prioritizing the state as the central unit of analysis, but also to assess interactions across and between different levels of governance, thereby providing a comprehensive picture of the levels of internal and nontraditional security in the Asia-Pacific.

Key Points

- Despite traditional understandings of security focusing on interstate relations, the internal security of a state in the Asia-Pacific is important to peace and stability across the region.
- Nontraditional security moves security beyond interstate military confrontation. It includes transnational challenges like climate change, which can cause mass displacement and in turn affect potential recipient countries like the United States.
- Nontraditional security refocuses security beyond the state as a unitary actor to include a web of complex interactions among different stakeholders, including communities, governments, international organizations, and nongovernmental organizations.
- Health, food, and water security are all priority concerns of the region, with a significant impact on the lives of millions and the region's economic development.
- Nontraditional and internal security issues are often highly interconnected, requiring multiple actors addressing several issues simultaneously to resolve.

Questions

1. What are the most significant internal security challenges that the Asia-Pacific states face?

2. How does history explain some of the internal security challenges that the Southeast Asian states face?

3. How do nontraditional security challenges differ from traditional ones?

4. Why do nontraditional security challenges require a comprehensive approach in order to deal with them?

5. Beyond the state, what institutions and organizations can help solve nontraditional and internal security challenges?

Guide for Further Reading

Acharya, Amitav. *Constructing a Security Community in Southeast Asia: ASEAN and the Problem of Regional Order.*London: Routledge, 2009.
This book provides a detailed explanation of regional cooperation and its limits. It includes investigation into transnational security issues such as SARS.

Caballero-Anthony, Mely, and Alistair D. B. Cook, eds. *Non-Traditional Security in Asia: Issues, Challenges and Framework for Action.* Singapore: Institute of Southeast Asian Studies, 2013.
This book provides an analytical framework to evaluate nontraditional security in the Asia-Pacific and provides case studies.

Capie, David, and Paul Evans. *The Asia-Pacific Security Lexicon.* Singapore: Institute of Southeast Asian Studies, 2002.
This book investigates the various terms used to address security issues in the Asia-Pacific.

Katzenstein, Peter, and Noburo Okawara. "Japan, Asian-Pacific Security and the Case for Analytical Eclecticism." *International Security* 26, no. 3 (2002): 153–85. http://dx.doi.org/10.1162/016228801753399754.
This article makes the case for focusing on problem-driven research drawing from multiple paradigms to explain real-world problems.

Tow, William, ed. *Security Politics in the Asia-Pacific: A Regional-Global Nexus.* New York: Cambridge University Press, 2008.
This edited book includes chapters on the security issues of the Asia-Pacific region, including diseases, energy, the environment, nuclear weapons, and terrorism. It demonstrates how many of the challenges facing the region are also global in nature and may require international cooperation to resolve.

Notes

1. Benjamin Reilly, "Internal Conflict and Regional Security in the Asia-Pacific," *Pacifica Review* 14, no. 1 (2002): 7–21.

2. Jemma Purdey, *Anti-Chinese Violence in Indonesia, 1996–1999* (Singapore: NUS Press, 2007).

3. David Carment, Patrick Hames, and Zeynep Taydas, "The Internationalisation of Ethnic Conflict: State, Society and Synthesis," *International Studies Review* 11 (2009): 77.

4. Mely Caballero-Anthony, "Revisiting the Bangsamoro Struggle: Contested Identities and Elusive Peace," *Asian Security* 3, no. 2 (2007): 141–61.

5. IPCC Secretariat, *Fourth Assessment Report: Climate Change 2007: Working Group I: The Physical Science Basis* (Geneva: IPCC Secretariat, 2007).

6. IPCC, "Summary for Policymakers," in *Climate Change 2013: The Physical Science Basis. Contribution of Working Group I to the Fifth Assessment Report of the Intergovernmental Panel on Climate Change,* ed. T. F. Stocker, D. Qin, G.-K. Plattner, M. Tignor, S. K. Allen, J. Boschung, A. Nauels, Y. Xia, V. Bex, and P.M. Midgley (Cambridge: Cambridge University Press, 2013), 11.

7. Patrick D. Nunn, "The End of the Pacific? Effects of Sea Level Rise on Pacific Island Livelihoods," *Singapore Journal of Tropical Geography* 34 (2013): 143–71.

8. EM-DAT, "Glossary," in *The International Disaster Database*, ed. EM-DAT (Brussels: EM-DAT, 2015).

9. Kristalina Georgieva, "Cost of Natural Disasters Has Quadrupled over Past 30 Years," Associated Press, June 5, 2014.

10. Institute for Environment and Human Security of the United Nations University, *World Risk Report* (Bonn: United Nations University, 2014), http://ehs.unu.edu/news/news/world-risk-report-2014.html#info.

11. Asian Development Bank, "Climate Change and Disaster Risk Reduction Assessment," in *Country Operations Business Plan: Philippines, 2013–2015* (Manila: Asian Development Bank, 2013).

12. White House, "Fact Sheet: US Response to Typhoon Haiyan," press release, November 19, 2013.

13. RSIS Centre for Non-Traditional Security Studies, *RSIS Non-Traditional Security (NTS) Year in Review 2014* (Singapore: RSIS Centre for Non-Traditional Security Studies, 2014), 16, http://www.rsis.edu.sg/wp-content/uploads/2014/12/NTS-YIR-2014.pdf.

14. Mely Caballero-Anthony, Alistair D. B. Cook, Belinda Chng, and Julie Balen, "Health Security," in *Non-Traditional Security in Asia: Issues, Challenges, and Framework for Action*, ed. Mely Caballero-Anthony and Alistair D. B. Cook (Singapore: Institute of Southeast Asian Studies, 2013), 17.

15. Kai Ostwald, "Ebola, SARS, and the Economies of Southeast Asia," *ISEAS Perspectives* 63 (2014): 4.

16. Ibid., 8.

17. RSIS Centre for Non-Traditional Security Studies, *RSIS Non-Traditional Security (NTS) Year in Review 2014*, 8.

18. Third APEC Ministerial Meeting on Food Security, "Beijing Declaration on APEC Food Security," September 19, 2014.

19. J. Jackson Ewing, "Supermarkets, Iron Buffalos, and Agrarian Myths: Exploring the Drivers and Impediments to Food Systems Modernization in Southeast Asia," *Pacific Review* 26, no. 5 (2013): 482–83.

20. Food and Agriculture Organization of the United Nations, *The State of Food Insecurity in the World 2013: The Multiple Dimensions of Food Security* (Rome: Food and Agriculture Organization of the United Nations, 2013).

21. Chheang Vannarith, "A Cambodian Perspective on Mekong River Water Security," *Stimson Analysis*, April 4, 2012, http://www.stimson.org/summaries/a-cambodian-on-mekong-river-water-security/.

22. X. X. Lu and R. Y. Siew, "Water Discharge and Sediment Flux Changes over the Past Decades in the Lower Mekong River: Possible Impacts of the Chinese Dams," *Hydrology and Earth System Sciences Discussions* 10, issue 2 (2006): 181–95.

23. "Lower Mekong Fish Production Estimated to Be Worth Up to $7 Bln a Year," *Cambodian Herald*, March 10, 2014.

24. Mekong River Commission, "Agreement on the Cooperation for the Sustainable Development of the Mekong River Basin," April 5,1995.

25. Cecilia Tortajada, Yugal Joshi,and Asit K. Biswas, *The Singapore Water Story: Sustainable Development in an Urban City-State* (Abingdon, UK: Routledge, 2013).

10 How Is the Cyber Revolution Changing Asia-Pacific National Security Concerns?

Rex B. Hughes

Cyberspace: A consensual hallucination experienced daily by billions of legitimate operators, in every nation, by children being taught mathematical concepts. . . . A graphic representation of data abstracted from the banks of every computer in the human system. Unthinkable complexity. Lines of light ranged in the nonspace of the mind, clusters and constellations of data. Like city lights, receding.

—William Gibson, *Neuromancer*

Reader's Guide

This chapter introduces you to the emergence of cyberspace as an Asia-Pacific national security concern. It outlines how strategic and tactical competition in cyberspace has become just as important as real space for determining the future security of the region. It begins by tracing the origins of cyberspace as a strategic security concern of the United States and NATO. It next examines the emergence of cyberspace as a twenty-first-century Asia-Pacific national security concern. It then examines the relationship of strategic cyberspace to the United States' pivot to the region. It concludes by reviewing the actions taken by Asia-Pacific great and middle power countries to secure their national cyber defense.

Introduction

This chapter discusses the increasing role of cyberspace in Asia-Pacific national security affairs, especially as it pertains to the United States' post-pivot role in the region. Thanks to the convergence of a complex array of technological, economic, demographic, linguistic, and cultural forces, cyberspace is becoming a central feature in Asia-Pacific national security affairs. Little did the American-Canadian William Gibson know when he popularized the term "cyberspace" in his 1984 steampunk book *Neuromancer* that cyberspace would become a defining element in twenty-first-century global security, particularly as it pertains to the high-growth Asia-Pacific.[1]

With the partial exception of constructivism, all the major security study theories discussed in this book were developed long before the global rise of the internet and cyberspace. As such, there is a significant debate over the use and merits of these theories in such a domain. Realists argue that cyber issues are just another form of human politics and so are subject to the same concerns of anarchy, a need for self-help, and power politics. However, their traditional focus on material strength (economics, military size) and hard power has led some to question the relevance in a global political economy driven by knowledge and innovation. Likewise, liberalism as a security studies theory is firmly grounded in material concerns, though it too struggles to address noninstitutional cooperation in an age of distributed network "crowdsourcing" and "creative commons" economic disruption.

Constructivism, with its emphasis on the role of ideas and perception, offers some useful insights (e.g., does it make sense to talk of a cyber "war"?[2]), but it also struggles to capture the impact of digital disruption enabled by a new constellation of nonstate actors and ideational ad hoc networks (as will be discussed below when looking at the cyber activities of the great powers in the Asia-Pacific). Perhaps this is an opportunity for critical security theories to make their mark by using this new environment to ask questions about our fundamental assumptions about how both state and nonstate cyber-enabled actors interact and whether this emergent digitally enabled ideational/institutional nexus can help avoid international conflict and encourage cooperation.

Cyberspace as a Primary National Security Concern

How did cyberspace evolve to become a primary twenty-first-century national security concern? While cyber terms such as **cybersecurity, cyber defense, cyberattack, cyberpower,** and **cyber deterrence** only entered the mainstream national security lexicon during the latter part of the last decade, the roots of cyberspace as a national security concern can be traced to the 1970s revolution in military affairs (table 10.1).

As the United States and its allies became increasingly obsessed with defeating the USSR without a direct military confrontation, a number of military planners and defense contractors set out to exploit the advanced information technologies and systems associated with the digital information revolution. Digital information technologies such as the microprocessor, integrated circuit, random access memory, advanced signal processing, global positioning system, and precision guidance technology were combined to create a new generation of information-centric smart weaponry. Advanced weapons platforms ranging from the M1A1 battle tank to the F-16 fighter jet were among the first generation of digital cyber-

Table 10.1 Key Cyber Terms

Term	Definition
Cyberattack	Offensive actions that target computer information systems and networks.
Cyber defense	Protective actions taken in anticipation of an attack against computer information systems and networks.
Cyber deterrence	Actions taken to dissuade adversaries from engaging in a cyberattack.
Cyberpower	The ability to use cyberspace to create advantages and influence events in other operational environments and across the instruments of power.
Cybersecurity	Measures relating to the confidentially, availability, and integrity of information that is processed, stored, and communicated by electronic or similar means.

enabled weapons systems built around an emergent information-centric military **doctrine**.

As more cybernetically controlled weapons proved their strategic and tactical value in the heat of battle, US military and political strategists began to craft a new information operations doctrine.[3] While early generations of these weapons were often unreliable and difficult to operate, the 1990 Gulf War was a watershed moment for the use of "smart weapons" on a dumb battlefield. As chronicled by CNN during the First Gulf War, smart weapons—conventional munitions guided by advanced information technology—tilted the battlefield toward adversaries with cybernetically remote-controlled weapons. Although they may not have realized at the time, the young operators of these digitally enabled smart weapons platforms were among the first generation of **information warriors.** As the decisive US win in the Gulf War showed, smart weaponry earned its position as the new currency of twenty-first-century warfare.

As smart weapon inventories expanded for the United States and its allies, the Pentagon and other national command headquarters soon required new techniques and strategies for coordinating and deploying these weapons, often over great distances in unknown environments. Thus, new computing and communications technologies would be needed to coordinate this new type of expeditionary force. While the US Department of Defense (DoD) was making incremental progress toward its goal of placing a computer on nearly every officer's desk, connecting these devices in a seamless global network was another matter. To solve this challenge, the DoD asked its Advanced Research Projects Agency (ARPA) to solve the digital command-and-control (C2) network issues. From the late 1960s onward, ARPA had channeled research toward a new generation of digital networking technology called "packet switching." Through this early digital research, the experimental "ARPANET" was born. More than just a way to connect the Pentagon's computers, ARPANET, through its innovative use of packet switched technology, was laying the groundwork for the Internet Age.

As the ARPANET and soon-to-be internet expanded starting in the early 1970s, a senior Boeing research scientist, Tom Rona, contemplated in his Seattle laboratory the long-term effects of the digital revolution on modern warfare. In a 1976 Boeing Research paper titled "Weapons Systems and Information Warfare," Rona speculated how the digital revolution would transform the full spectrum of battlefield weapons, ranging from space to the maritime domain. In order to take advantage of Rona's vision, the DoD would need to accelerate its research and development and the integration of digital technologies into its future weapons systems, especially if it wanted to increase the United States / USSR technology gap. Rona made doctrinal contributions to the US information revolution in military affairs, as the father of **information warfare**. He also played a key role in creating the largest software company in the world, for he used Boeing research funds to sponsor some of the University of Washington's computers on which Bill Gates and Paul Allen learned to program. Thus, for realists, whereas cyberpower might be "virtual," a more thorough understanding of the role hard power played in the development of the United States–led global computing industry might challenge the traditional realist and liberal internationalist cleavage of soft power and hard power concepts.[4]

As the Gates and Allen Microsoft-IBM personal computer (PC) revolution took hold during the 1980s, DoD joined their quest for greater global connectivity and information exchange. By the early 1990s, after a number of failed attempts to network the world's PCs, ARPANET's internet proved the best candidate for the job for both civilian and military networks. However, connecting PCs to the internet meant that PCs needed to be much more resilient against a growing number of computer viruses and cyberattacks. While most of these early attacks on connected PCs were carried out by pranksters and pimply teenagers, that changed substantially with the advent of global electronic commerce in the late 1990s. Thanks to the convergence of the PC and internet revolutions, cyberspace was well on the way to becoming the information domain of the twenty-first-century global economy.

Cyberattack as a National Security Concern

By the turn of the century, DoD war planners were awakening to the broadening scale and scope of cyberattacks, thanks in part to the success of the internet around the globe. By the late 1980s, computer viruses such as Animal, Jerusalem, and Ghostball were routinely penetrating DoD computers. Rona's worst nightmare became a real threat as some strategic adversaries were also beginning to use information as a weapon, thus turning the US edge in advanced information technology into a strategic vulnerability.

Despite the decades-long expansion of military digital computing and communications, it was not really until the 2007–10 period that cybersecurity began to emerge as a major national security concern for the United States and its allies. In 2007, the Estonian government triggered what will likely go down in history as the first multinational cyber defense emergency. At the time, what seemed an innocuous decision made by the Estonian Ministry of Defense to move a bronze Soviet soldier statue from central Tallinn Square to a military base resulted in a sustained campaign of military-grade denial of service against its government and commercial information infrastructure. As the attacks escalated beyond what the Estonian authorities could defend on their own, the North Atlantic Council of NATO was convened to explore what aid could be offered. After NATO discovered that it had few deployable cyber defense assets (human or machine) that could be offered to Tallinn, the United States and other alliance members decided that it was time to act against future cyber threats in a more unified way.

A year after the Estonian attack, and after DoD suffered its own worst cyberattack—"Operation Buckshot Yankee"—DoD formed the world's first Cyber Command, USCYBERCOM. With the establishment of USCYBERCOM, the United States declared cyberspace to be a major area of US national security concern. USCYBERCOM also gave legitimacy to cyberspace becoming an active battle space, whether in peacetime or war. Other nations would soon follow with their own militarization of cyberspace capabilities, including those in the Asia-Pacific.

Cyberspace as an Asia-Pacific National Security Concern

While there is no single watershed Asia-Pacific cyber defense event comparable to the Estonian bronze soldier incident, there have been several high-profile cyberattacks since 2005 that have helped elevate cyberspace to a regional national security concern. Only during the last five years has cyberspace emerged as an issue of strategic importance for the region's dominant economic and military powers. To date, China, Russia, Japan, Australia, South Korea, and North Korea have been the most consequential cyber actors in the Asia-Pacific region. Over time, other smaller states are expected to increase their cyber capabilities. As is the case with real space, the US-China competition is today—and likely to be for the foreseeable future—the key driver of Asia-Pacific strategic competition in cyberspace. The United States and China have already demonstrated their willingness to project national power to achieve strategic objectives in cyberspace. Russia is not far behind China in its capabilities, but its own Pacific pivot only started in 2014 following its China energy deal and Crimea annexation (table 10.2).

Beginning as early as 2005, the US government publicly acknowledged a series of sustained, advanced, persistent attacks directed against US federal government

Table 10.2 Recent International Cyberattacks

State	Date	Type
United States	2003–5	"Titan Rain" attacks on government targets. Likely sponsored by China.
China	October 2007	Alleged attack on aerospace and other sensitive industries from Taiwan and the United States.
South Korea, United States	July 2009	Denial-of-service attacks. Likely sponsored by North Korea.
Australia	October 2010	Large increase in attacks against Defence Ministry targets reported.
Japan	August 2011	Information stolen from defense-linked corporations.
Japan	September 2012	Increase in attacks on Japanese government sites coinciding with Senkaku/Diaoyu nationalization controversy.
United States	2013	Leaks reveal National Security Agency activities against China, Russia, North Korea, and Group of Twenty members, including mass and targeted surveillance.
China	December 2013	China's Central Bank disrupted by Bitcoin users in protest against regulations.

Source: Center for Strategic and International Studies, *Significant Cyber Incidents since 2006* (Washington, DC: Center for Strategic and International Studies, 2014), http://csis.org/files/publication/140310_Significant_Cyber_Incidents_Since_2006.pdf.

computer systems—thought to have been carried out by China's People's Liberation Army (PLA). The code name given to the attacks by the US government was operation Titan Rain, which is thought to have run since 2003. The attacks resulted in the first formal DoD strategic assessment of the PLA as a state sponsor of cyberattacks. When confronted by US officials, the Chinese government denied all claims. In October 2007 the Chinese government publicly retaliated against the US accusations by accusing the United States of stealing sensitive information from state-sponsored enterprises, such as the China Aerospace Science and Industry Corporation. At the time, US officials brushed off these Chinese government accusations.

The 2008 Mumbai terrorist attack was a key watershed event for India and Southeast Asia regional cybersecurity.[5] On November 26, 2008, members of the Lashkar-e-Taiba terrorist group attacked the hotel and retail core of South Mumbai. Unknown to Indian Intelligence officers at the time, the Mumbai attackers made sophisticated use of internet and mobile technologies such as VOIP (Voice Over Internet Protocol), instant messaging, and global positioning system mapping to plan and coordinate their attacks. Subsequent reviews showed substantial weaknesses in the capacity of India's intelligence agencies' ability to map the group's electronic footprints leading to the attack. Subsequent forensic analysis conducted by the US Federal Bureau of Investigation (FBI) on behalf of Indian investigating teams showed that the attackers routed their cellular calls over the internet via the United States in order to avoid detection by the security services. This attack was a game changer for Asia-Pacific state security services because it clearly demonstrated the ability of nonstate actors to escape conventional analogue telecommunications surveillance.

Even East Asia's most cyber-enabled state has suffered crippling attacks. In July 2009 major portions of South Korea's internet came under a massive denial-of-service attack. Organizations including the country's largest bank, telecommunications provider, and newspaper websites all suffered crippling attacks to their online services.[6] The internal and external digital networks of US forces in South Korea were also affected by the massive botnet, which encompassed over 100,000 hosts. While no specific nation or attacker could be identified as leading the attack, given that the activity was directed at South Korea's host computers, there was much speculation that a newly formed North Korean cyber military Unit 121 was behind the attack and also the recent high-profile 2014 Sony Pictures attack.[7] In the ensuing years, similar attacks have been carried out against South Korean targets. In March 2015 the South Korean government blamed North Korea for cyberattacks on its Korea Hydro and Nuclear Power Co. Ltd. plants in late 2014.[8]

In 2011 Japan suffered a major national embarrassment when leading Japanese defense contractor Mitsubishi Heavy confirmed reports that its internal network had been breached and sensitive information related to the US Joint Strike Fighter (JSF) had been successfully exfiltrated. The United States reacted very negatively to this event and was critical of Japan's state of cybersecurity readiness. Boeing, the company leading the JSF project, complained of possible intellectual property theft related to its JSF work. Later, suspicions arose that the Chinese government was behind the attacks and that Mitsubishi's cybersecurity was systemically weak and out of step with industry best practices. The attack served as a wake-up call for the Japanese government. Other attacks against government targets—including those of the Ministry of Foreign Affairs, Japan Self-Defense Forces, and the Ministry of International Trade and Industry—had also been revealed (some of which occurred since 1998), with those in 2011 deemed Japan's first "cyberwar."[9]

In the spring of 2012, a range of advanced cyberattacks were directed against high-profile Japanese websites after a territorial dispute with China over the Senkaku-Diaoyu Islands. In September 2012, the Japanese government reported a substantial increase in cyberattacks against high-profile government websites, following the government's move to buy three islands from private Japanese owners. Japan's National Police Agency announced that three hundred sites were identified as targets on the Chinese **hacktivist** website Honker Union.[10] While these attacks eventually died down, they served as a stark warning and motivation for Japan to invest more earnestly in its nascent national cyber forces.

The most disruptive and damaging cyber intelligence incident for US Asia-Pacific relations was the 2013–14 National Security Agency (NSA) revelations by a former employee, Edward Snowden. For decades, many Asia-Pacific countries harbored suspicions about the scale and scope of NSA surveillance. In late 2013 many of these suspicions were shown to be true, after Snowden flew covertly to Hong Kong and began leaking information to *The Guardian* and other high-profile news outlets. In the ensuing months, following Snowden's Moscow asylum, much of the ultrasecret NSA spycraft was made public. Many Asia-Pacific nations publicly protested NSA surveillance.[11] Perhaps most significantly, this incident created uniquely widespread incentives for Asia-Pacific citizens and nations to use advanced cryptography and other mass surveillance-thwarting techniques to hide their communications from NSA's snooping. As constructivists have pointed out, one important feature of a new transnational domain's emergence is its ability to challenge prevailing norms, including those for vital issues of national security such as the role and scope of the state's intelligence services. Since the Edward Snowden NSA disclosures, many Asia-Pacific states have upgraded their antisurveillance systems and policies as well as bilateral cybersecurity cooperation with the United States, including close allies such as Australia. Strategic US adversaries such as China and Russia have been particularly vocal in calling the United States to task for so-called human rights violations and antidemocratic policies carried out by the NSA in its global surveillance programs.[12] Some experts have even gone as far as to speculate that the Snowden NSA revelations could lead to a situation where the internet is Balkanized into spheres of influence, especially if China and other Asia-Pacific nations begin building alternative closed-end networks.[13]

In a highly publicized, precedent-setting move by the United States, on May 19, 2014, a grand jury in the Western District of Pennsylvania indicted five Chinese military hackers for computer hacking, economic espionage, and other offenses directed at six American high-value targets in the US nuclear power, materials, and solar products industries. The accusation was significant because it marked the first time that the United States had leveled criminal charges against senior PLA military officers. The decision by the US Department of Justice to bring these unusual charges against the PLA was apparently spurred by the Chinese government's request for hard evidence related to past charges, or what former NSA director General Keith Alexander referenced as "the greatest theft of intellectual property in history."[14] While it is uncertain where these charges are heading and what long-term damage they may cause to US-China relations, there is already evidence of lasting harm. China has retaliated with a series of trading bans against the US high-technology firms IBM and Microsoft, and also against the management consulting firm McKinsey & Company.

Cyberspace and the Great Powers of the Asia-Pacific

This section explores how the major states of the Asia-Pacific are expanding into the cyber domain and using it to further their strategic interests. It focuses particularly on the United States, which is not only one of the most active participants but also the most advanced in its technology and capacity. This makes it an important case study for understanding the likely direction of other states.

Since the United States' pivot to the Asia-Pacific was announced by President Barack Obama in November 2011, the country has taken a series of economic, diplomatic, and military steps to exert its influence on the strategic direction of Asia-Pacific cyberspace. At the grand strategic level, US national security objectives in Asia-Pacific cyberspace are similar to its strategic goals in the wider Asia-Pacific. As of 2015, there are at least seven identifiable objectives for the cyber dimension of the United States' regional rebalance.

First and foremost, the United States seeks to maintain a stable, secure, and open internet in the Asia-Pacific. Even before the pivot, US officials from multiple civilian and military agencies had been quietly working behind the scenes to strengthen the region's internet infrastructure and governance mechanisms. In the Asia-Pacific, a United States–led regional and national computer emergency response team (CERT) model has been adopted, whereby each state has a national CERT which provides the single point of contact for cybersecurity issues affecting its businesses. The national CERTs then coordinate their activities through the regional Asia-Pacific CERT (APCERT). The United States has also worked behind the scenes to see that nations adopt common open internet standards according to guidelines from the Internet Engineering Task Force and the International Organization for Standardization. The United States has offered technical assistance to countries seeking to upgrade their internet infrastructure and to deploy the latest internet protocol software and routing techniques. Several US government agencies, such as the State Department and the Voice of America—a broadcaster funded by the federal government—are involved in internet anticensorship campaigns. A number of key technologies have been developed by the Voice of America and deployed in the region.[15] Liberal international relations scholars suggest that because there are many common challenges facing all nations—along with a widespread lack of clear principles, rules, and norms about the domain—it should be easier for states to find paths for positive-sum online cooperation.

Second, as a response to accusations by China and Russia that it is an imperialist internet hegemon, the United States has sought to establish a common "code of conduct" or "rules of the road" for cyberspace. The origins of this diplomatic conversation can be traced to the 2003 United Nations World Summit on Information Society. At this time, the UN established the permanent Group of Governmental Experts (GGE) to tackle the issue of cyberspace norms. The Asia-Pacific GGE members are Australia, Canada, China, India, Indonesia, Japan, Russia, and the United States.[16] The most contentious area of discussion, especially with respect to the regional interests of China and Russia, is the national information security dialogues. Whereas China and Russia both seek a GGE consensus on the use of so-called information controls to prevent internal or external descent, the United States and other liberal democratic members firmly reject such controls. These discussions, already difficult, have become much more arduous since the Snowden NSA revelations, with even countries such as Brazil and Germany raising voices of dissent from the US position. If a satisfactory consensus cannot be reached in the GGE, the United States is likely to turn to regional bodies such as the Asia-Pacific Economic Cooperation (APEC) and the ASEAN Regional Forum for rules-of-the-road discussions, especially for cybersecurity issues.

Third, although no major global or regional armed conflict is on the immediate horizon, DoD continues to place a high priority on winning decisively in any interstate armed conflict, including one in cyberspace. As chronicled in the beginning of this chapter, US strategic thinking has evolved to see cyberspace as a military domain equivalent to those of land, sea, air, and space. In practical terms, this means that DoD is investing serious resources in developing not only robust cyber defenses but also a vast array of cyberattack / active defense technologies.[17] At this time it is unclear how far the United States is willing to go with its active defense capabilities to defeat a state or nonstate opponent in cyberspace; some US military leaders and experts have indicated that the United States reserves the right to even use nuclear weapons should a cyberattack result in a catastrophic loss of life. The Snowden disclosures have revealed that the United States devotes a significant amount of advanced cyber spycraft to global surveillance and signals intelligence, as well as the strategic importance of the Asia-Pacific to US economic, diplomatic, and military interests. It can therefore be expected that the United States will expend a serious amount of treasure to win any protracted conflict in cyberspace. As yet it is unclear if Russia or China would do the same.

Fourth, since 2008 the US government has expressed concern about the rising level of cybercrime in the Asia-Pacific. Due to the growth of cybercrime in countries such as China, Malaysia, and Indonesia, US law enforcement agencies have devoted substantially more resources to tracking cybercriminals from the region. The United States is a signatory to the 2001 Budapest Convention on Cybercrime, and is in the process of assisting Asia-Pacific signatories to implement recommended procedures, especially with respect to processing evidence and criminal prosecutions. The US government is also devoting more human and technical resources to its federal law enforcement officers charged with prosecuting overseas cybercrime such as the FBI and Secret Service. The US government is also an active supporter and participant in the Interpol Cyber Innovation Centre in Singapore.[18]

Fifth, China is viewed by the United States as its main strategic competitor in cyberspace, with Russia a close second. China's increasing investment in cyber defense and cyber-enabled kinetic weapons platforms worries US war planners. US-China cooperation on cyberspace has been part of the US–China Strategic and Economic Dialogue since 2009. On the economic side of the discussions, the United States has expressed its own distress with China's perceived lack of enforcement of digital intellectual property protections and trade secrets theft. On the strategic side, US officials seek to press China to establish rules of the road for information security disputes. Ironically, the Snowden revelations occurred during the 2013 Summit in Sunnylands, California, between US president Barack Obama and Chinese president Xi Jinping. In the subsequent weeks, the United States protested China's refusal to surrender Snowden to the US authorities in Hong Kong.

Sixth, as in other areas of its Asia-Pacific multilateral relations, the US government currently places a high priority on its regional and bilateral alliances. In keeping with its 2015 maritime "Cooperative Strategy for 21st Century Sea Power," the United States also seeks multilateral cyber contributions—both in critical areas of regional concern and in situations where US cyber defense capabilities may be depleted or stretched thin. The United States is also in the process of reaching out to key alliance members to reassure them that the United States "has their back" should China extend its recent South China Sea / East China Sea provocation—as discussed in chapter 7—to the cyber domain. However, the US government is also investing a substantial amount of **diplomacy** in repairing the damage done to its Asia-Pacific bilateral alliances by the Snowden revelations.

Multilateral organizations and alliances also play an important role in the US Asia-Pacific cyber engagement. APEC and the ASEAN Regional Forum are two of the Asia-Pacific economic and security multilateral organizations in which the United States participates in cyber engagements. In APEC, the United States supports a number of cyber-relevant working groups, such as Telecommunications and Transportation. In the ASEAN Regional Forum, the United States has been supportive of a number of cyber initiatives, ranging from confidence-building measures to cyber incident response to critical infrastructure. These United States–led diplomatic regional engagements are expected to deepen in the coming years.

Seventh and finally, the United States is extensively integrating cyber capacities into its Asia-Pacific military strategy, doctrine, and tactics. An important example of this is Washington's application of cyberpower to its evolving Air-Sea Battle (ASB) concept. Given China's rapidly growing blue water navy and its burgeoning anti-access / area denial (A2/AD) capabilities, the United States is expected to devote substantially more resources to integrating cyberpower with its ASB strategy, even though at the time of writing it is in the process of reconsidering the name for and bureaucratic approach to the concept.[19] While DoD has stayed largely tight-lipped on the cyber component of its ASB strategy, cyberpower is expected to play a greater role in ASB tactics as opposing naval fleets and air platforms take on new information processing and network-enabled capabilities.[20]

For the United States and its Asia-Pacific partners, China's A2/AD expansion will likely require a new ASB cyber doctrine to manage the threats posed by a new generation of network-enabled platforms. Because the United States is far advanced in the integration of cyber with ASB capabilities, there is a risk for China in deploying a next-generation, network-enabled force, especially if the United States can leverage Cyber Command / NSA know-how to exploit or embed terminal defects or firmware backdoors. Whereas traditional electronic warfare jamming has constituted the main offensive component of Asia-Pacific ASB US Naval assets, such as the F-18 Growler electronic warfare fighter-bomber and the new P-8 Poseidon Multimission Maritime Airfraft, China's long-term migration to smarter network-enabled platforms, such as its J-31 fifth-generation fighter, will likely require these platforms to more closely integrate cyberattack capabilities with traditional electronic warfare. If the United States commits to linking cyberattack to electronic warfare capabilities on these and other platforms, China may need to reconsider its long-term A2/AD strategy inside the China Sea nine-dash line. Given its advanced technological base and participation in codeveloping advanced commercial network-enabled platforms such as the Boeing 787 Dreamliner, Japan could emerge as the most critical US regional strategic partner in the development of next-generation, network-enabled platforms for United States–led ASB 2.0 platforms.[21]

The Asia-Pacific Cyber Great Powers

The United States is far from the only country making cyber technology a major focus in its approach to the Asia-Pacific. China is the largest and fastest-growing cyber economy both in the region and in the world. As of January 2014, the China Internet Network Information Center reported that there are 627 million internet users in China. Since the late 1990s China has developed the full complement of institutions necessary to manage its growing regional and global presence in cyberspace. China has an internet-savvy Ministry of Information responsible for regulating and censoring Chinese cyberspace, also known as the "Great Firewall." As in other parts of the country, the PLA plays a substantial role in China's foreign and domestic cyber affairs. The United States and its Asia-Pacific allies have identified a range of PLA-sponsored cyber economy businesses, ranging from network equipment to consumer software. Japan has accused PLA information operations units such as the infamous PLA Unit 61398 of trying to manipulate public opinion away from Japan and toward China in the Senkaku/Diaoyu Islands dispute. In global diplomatic discussions about the "rules of the road" for cyberspace, China has aligned with Russia against the US and EU position in their GGE cyber norms engagement.[22]

Japan is the third-largest economy globally and the second-largest cyber economy in the Asia-Pacific region. Given its technological edge in most information and communications technology categories, Japan is and will continue to be a great cyberpower for the foreseeable future. Japan has a well-developed internet governance infrastructure and established one of the region's earliest national CERTs. Tokyo is also home to the regional APCERT.

The US-Japan relationship is the cornerstone of the United States' rebalance to the region and the United States' Asia-Pacific cyber defense, with the White House observing that "the United States and Japan share a commitment to an open, interoperable, secure, and reliable cyberspace."[23] In recent years there has been a series of high level bilateral discussions about cooperation on cybersecurity. A Cyber Defense Policy Working Group and biannual dialogues are coordinated jointly by the Japanese Ministry of Defense and DoD.[24] A major US-Japan cyberwarfare exercise between the US Army and Japanese Self-Defense Forces was held in Hokaido in December 2013 to coordinate cyber defense in a wartime situation.[25]

The Japanese Self-Defense Forces have significant cyber defense and attack capabilities. Thanks to an expanded US-Japan defense cooperation agreement signed in 2013, Japan is in the process of integrating a growing number of its national cyber defenses with the United States. In March 2014 the Japanese Self-Defense Forces established a Cyber Defense Unit. Although the Japanese Constitution forbids Japan to attack other nations, Japan is still thought to harbor advanced cyberattack capabilities if needed. As part of the United States' pivot to the Asia-Pacific, the United States is already seen as seeking cooperation from Japan and its cyber capabilities.

India is the third-largest Asia-Pacific cyber economy, with an estimated population of 237 million internet users. The Indian economy is heavily dependent on information and communications technology, with the bulk of its high-technology jobs coming from software services and support. However, despite its service economy centered on information and communications technology, India's internet users make up less than 20 percent of the population and the cyber economy accounts for less than 2 percent of gross domestic product. India's internet governance infrastructure is well developed. The Government of India established a national CERT in 2004 and is a member of APCERT. The Indian government has also indicated that cyber capabilities will be a major investment area in its defense modernization efforts. Lashkar-e-Taiba's successful 2008 use of the internet to coordinate the Mumbai attack serves as a stark reminder of the gap the Indian government has with potential rivals. Indian defense officials have expressed concern about China's growing cyberattack capabilities. The use of cyberspace by Pakistan and tribal territory groups is also of concern to India.

Given India's deep relationships with the US high-technology industry, it is likely that India will increasingly seek cooperative cyber defense ties with the United States. India already participates in the annual US Cyber Storm exercise as an observer. However, despite close cooperation and expanding trade in advanced information and communications technology software services, the United States and India have not advanced far together in their bilateral cyber relations. Since 2010 the two countries have engaged in an annual strategic cyber dialogue on issues ranging from cooperation on cybercrime to reducing trade barriers. However, no significant initiatives have emerged, and the relationship lacks the strategic and operational depth of the US-Japan or the US-China cyber relations. Given the 2014 sweeping victory of the Bharatiya Janata Party of India president Narendra Modi and the enormous role played by social media, it is expected that US-India bilateral cyber relations will see renewed strengthening in the coming years.

Despite its reliance on natural resources, Russia is among the top cyberpowers in the world. A growing portion of its economy relies on the internet for trade, and its military and security organizations contain a full spectrum of offensive and defensive cyber capabilities. Some of these capabilities were demonstrated during the Russian incursions into South Ossetia in 2008 and into Crimea in 2013.[26] During both these incidents, Russian forces seized key internet chokepoints by cyber means. As Russia ties more of its economic future to energy trade with China, it can be expected that its national security apparatus will devote more attention and resources to monitoring and shaping events in the Asia-Pacific strategic cyberspace (table 10.3).

Table 10.3 Internet Usage in the Major Asia-Pacific States

State	Number of Internet Users
China	627 million
United States	277 million
India	237 million
Japan	109 million
South Korea	45 million
Indonesia	42 million
Vietnam	40 million
Australia	18 million

Source: Central Intelligence Agency, *Country Comparison: Internet Users, CIA World Factbook 2014* (Washington, DC: Central Intelligence Agency, 2014), https://cia.gov/library/publications/the-world-factbook/rankorder/2153rank.html.

The Asia-Pacific Cyber Middle Powers

South Korea is widely considered one of the most advanced Asia-Pacific cyber economies, thanks to its deep broadband diffusion, dominance in mobile devices, and mature cyber governance infrastructure. South Korea has a National Cyber Security Strategy Council, which directs the development of national cybersecurity infrastructure and policy coordination across multiple agencies. The National Cyber Security Center is in charge of threat management. South Korea also has agencies devoted to cyber warfare and intelligence. In 2011 South Korea announced a National Cyber Security Strategy. South Korea also established KN-CERT in 2004, and is a member of APCERT. South Korea is capable of offensive cyber warfare, which is largely directed at North Korea. South Korea is the first Asia-Pacific country to emulate the US model for Cyber Command. Some have speculated that this new Cyber Command will be used to derail North Korea's nuclear weapons program covertly via cyber means.[27]

Indonesia is a developing cyber economy with 40 million internet users. While Indonesia does practice internet censorship, there are some signs of improvement or lessening controls. Indonesia is an active participant in regional governance organizations engaged in cyber affairs, such as APEC and ASEAN. It is a member of the GGE, has a national CERT, and is a member of APCERT. In 2012 the Indonesian minister of defense announced plans to create a Cyber Defense Operations Center for the coordination of national cyber defenses. Indonesia does not harbor the ability to conduct serious cyber offensive operations against known aggressors. However, the designated terrorist organizations Gerakan Aceh Merdeka, Organisasi Papua Merdeka, and the Al Qaeda affiliate Jemaah Islamiyah are all thought to be using cyberspace as a communications and recruitment tool within Indonesia.

Although North Korea has no notable cyber economy and boasts one of the lowest internet penetrations in the world, it is thought to have some cyberattack capabilities to wreak mischief on the internet. There is speculation that the General Staff Reconnaissance Bureau is the main driver of its cyber defense and cyberattack capabilities. There is much speculation in the cybersecurity community globally that "Unit 121" is responsible for North Korean attack capabilities. South Korea is thought to bear the brunt of Unit 121 attacks. North Korean supreme leader Kim Jong-un is thought to be an avid internet user and to be interested in North Korea's strategic exploitation of cyberspace. North Korea does not have a national CERT; nor does it participate in any regional or relevant international bodies.

With an internet population of 18 million, Australia is a middle power making outsized investments in its national cyber defenses. As a member of the five-eye intelligence club, Australia plays a major role in cooperative intelligence collection with the United States, the United Kingdom, Canada, and New Zealand.

Recognizing China as a major cyber threat, in 2010 Australia launched a series of initiatives largely based on those from the UK Office of Cyber Security and Information Assurance.[28] At the initial formation announcement of this new office, the then–prime minister, Julia Gillard, stated that it would become the hub of the government's cybersecurity efforts.[29] Australia launched a new Cyber Security Centre (ACSC) in June 2014. According to the Australian Department of Defence, the ACSC is designed to be a one-stop shop for cybersecurity operations by the Australian Signals Directorate, Defence Intelligence Organisation, Australian Security Intelligence Organisation, the Attorney-General Department's CERT, the Australian Federal Police, and the Australian Crime Commission. The official policy of the Australian government is to condone internet censorship, both foreign and domestic. Australia CERT is a member of APCERT.

Due to its five-eyes status and deepening defense cooperation with United States, Australia has significant offensive and defensive cyber defense capabilities. The Australian Signals Directorate administers the Cyber Security Operations Centre for the Ministry of Defence. Australia is highly reliant on the internet for economic activity, which makes it highly sensitive to cyberattack.

Singapore is a highly advanced cyber economy. In keeping with Singapore's penchant for well-executed central planning, the country is guided by a National Cyber Security Master Plan, which was launched in 2013. Singapore is also in the process of building a National Cyber Security Centre, which is coordinated by INFOCOMM and has passed legislation designed to curb cybercrime. With one of the oldest Asia-Pacific CERTs—SingCERT—Singapore is a member of APCERT. Unique to Singapore is its evolving trusted hub concept that builds on its strategic position as host to several multinational firms. Approximately 80 percent of Singapore users have access to mobile or broadband internet. Singapore is highly integrated with Malaysian cyberspace. It does practice some online censorship, although trends show that this is decreasing. Singapore is also a cyber-influence center because it has offices of a number of multinational high-technology companies such as Microsoft and Google, as well as regional governance centers such as APEC and the Interpol Cyber Innovation Center.

Conclusion

Twenty-first-century cyberspace is more than Gibsonian steampunk science fiction or a global geek playground. Due to the growing economic, political, social, and cultural reliance on cyberspace, it is an essential national security space. As this chapter has chronicled, cyberspace will increasingly feature in the United States' pivot to the Asia-Pacific. As recent US-China Strategic and Economic Dialogues have shown, both nations place a high priority on defending their national interests and sovereignty in cyberspace. Japan, India, Russia, and other great Asia-Pacific cyberpowers also seek competitive advantage in cyberspace.

Since the early 1970s, the United States has sought to exploit new information technologies for its strategic advantage—but China and the United States' other Asia-Pacific rivals have only done so in the last decade or so. While the United States continues to hold a decisive information advantage in most strategic domains, as the Snowden revelations have demonstrated, US cyber hegemony remains vulnerable to leaks and traditional spycraft. As the Asia-Pacific grows at historic rates and China's economic and strategic influence continues to rise, the ability of the United States to shape the communications architecture of Asia-Pacific cyberspace may indeed decline relatively over time. As the erstwhile software pioneer Microsoft has experienced with the surprising rise of its archcompetitor Google, the commanding heights of cyberspace can be lost at internet speed. Should China or other competitor nations attempt to overly restrict US access or freedom of movement to Asia-Pacific cyberspace, regional prosperity and growth could be threatened. Cyberpower could be a decisive factor in future contested zones such as the South China Sea and the East China Sea. Given the United States' absolute advantage in next-generation, network-enabled platforms such as the F-35 fighter and Ageis cruisers, potential regional adversaries such as China must proceed with caution in developing its own indigenous network-enabled cyber technologies and platforms. While many aspects of cyberpower inside the US Pacific pivot are still works in progress, there should be little doubt that the United States will seek to use its information advantage to shape the digital rules of the game for Asia-Pacific strategic cyberspace for decades to come.

Key Points

- Cyberspace has now become as important as the air, land, sea, and space domains for determining the future security of the Asia-Pacific.
- Cyber weapons have become increasingly efficient and accurate, making cyber-driven smart weaponry the new must-have equipment of twenty-first-century warfare.
- Despite the decades'-long expansion of military digital computing and communications, it was not really until 2007 that cybersecurity began to emerge as a major national security concern.
- In the Asia-Pacific, there have been several high-profile cyberattack events since 2005 that have driven cyberspace onto the national security radar screens of the region's dominant economic and military powers, as well as middle powers and small states.
- Cyberspace is a critical component of the US Pacific pivot and a key dimension of US-China strategic competition in the Asia-Pacific.
- The Asia-Pacific states have taken both individual and coordinated responses to the growing number of cybersecurity threats.

Questions

1. What differentiates cyberspace from other strategic environments such as land, sea, air, and space?

2. What role can cyber technology play in future warfare and weaponry?

3. How is cyberspace a primary security concern for Asia-Pacific states?

4. How are Asia-Pacific states responding to cybersecurity threats?

5. Does cyber technology "level the playing field" by making it easier for smaller states and nonstate actors to challenge larger states?

Guide to Further Reading

Ball, Desmond, and Gary Waters. "Cyber Defense and Warfare." *Security Challenges* 9 (2013): 91–98.
   This article considers the steps that Australia is taking to protect against cyberattacks and makes recommendations for how cybersecurity could be pursued in Australia and elsewhere in the future.

Farwell, James P., and Rafal Rohozinski. "Stuxnet and the Future of Cyber War." *Survival* 53, no. 1 (2011): 23–40. http://dx.doi.org/10.1080/00396338.2011.555586.
   This article uses the attack by the Stuxnet cyberworm on an Iranian nuclear facility as a case study to analyze how cyber technology can be used as weapons of war.

Hansen, Lene, and Helen Nissenbaum. "Digital Disaster, Cyber Security, and the Copenhagen School." *International Studies Quarterly* 53, no. 4 (2009): 1155–75. http://dx.doi.org/10.1111/j.1468-2478.2009.00572.x.
   This article uses the Copenhagen School's concept of "securitization" to analyze cybersecurity as a distinct sector of security that has particular threats and referent objects.

Rid, Thomas. *Cyber War Will Not Take Place*. London: Hurst / Oxford University Press, 2013.
   This book argues that cyberwar is likely to have less impact than nonviolent threats in cyberspace, including espionage, sabotage, and subversion.

Singer, Peter, and Allan Friedman. *Cybersecurity and Cyberwar: What Everyone Needs to Know*. New York: Oxford University Press, 2014.
   This book provides excellent background on how cyberspace works and what security challenges it may pose.

Notes

1. William Gibson, *Neuromancer* (New York: Ace Books, 1984).

2. Thomas Rid, *Cyber War Will Not Take Place* (London: Hurst / Oxford University Press 2013).

3. For a deeper understanding of cybernetic control theory, see Norbert Wiener, *Cybernetics; or, Control and Communication in the Animal and the Machine* (Cambridge, MA: MIT Press, 1961).

4. Kenneth Flamm, *Creating the Computer: Government, Industry, and High Technology* (Washington, DC: Brookings Institution Press, 1988), 90–120.

5. Raghu Santanam, M. Sethumadhavan, and Mohit Virendra, *Cyber Security, Cyber Crime and Cyber Forensics: Applications and Perspectives* (Hershey, PA: Information Science Reference, 2011), 49, http://my.safaribooksonline.com/book/-/9781609601232 /chapter-4-emergency-response-to-mumbai-terror-attacks/mumbai_terror_attacks _2611#X2ludGVybmFsX0J2ZGVwRmxhc2hSZWFkZXI/eG1saWWQ9OTc4MTYwOTY wMTIzMi80OQ==.

6. Matthew Weaver, "Cyber Attackers Target South Korea and US," *The Guardian*, July 8, 2009, http://www.theguardian.com/world/2009/jul/08/south-korea-cyber-attack.

7. Samuel Gibbs, "Did North Korea's Notorious Unit 121 Cyber Army Hack Sony Pictures?" *The Guardian*, December 2, 2014, http://www.theguardian.com/technology/2014 /dec/02/north-korea-hack-sony-pictures-brad-pitt-fury.

8. Prosecutors said the cyberattacks were made on December 9–12 via 5,986 phishing e-mails containing malicious codes that were sent to 3,571 employees of Korea Nuclear Power Ltd. Ju-min Park and Meeyoung Cho, "South Korea Blames North Korea for December Hack on Nuclear Operator," Reuters, March 17, 2015, http://www.reuters.com/article /2015/03/17/us-nuclear-southkorea-northkorea-idUSKBN0MD0GR20150317.

9. Jun Osawa, "Japan Faces New Cyberwar Ear," Institute for International Policy Studies, http://www.iips.org/en/research/data/oosawa01.pdf.

10. "Chinese Cyber Attacks Hit Japan over Islands Dispute," *Globe and Mail*, September 19, 2012, http://www.theglobeandmail.com/news/world/chinese-cyber-attacks-hit -japan-over-islands-dispute/article4553048/.

11. "Australians Urged to Protest over NSA Surveillance," *The Age*, February 12, 2014, www.theage.com.au/it-pro/business-it/australians-urged-to-support-protest-over-nsa -surveillance-20140212-hvc28.html.

12. "China Issues Report on US Human Rights," Xinhuanet, http://news.xinhuanet .com/english/china/2014-02/28/c_133149412.htm.

13. "Warning over Fragmented Internet," *Financial Times*, July 3, 2014, http://www.ft .com/intl/cms/s/0/a8d6e4ce-02d0-11e4-81b1-00144feab7de.html.

14. Josh Rogin, "NSA Chief: Cybercrime Constitutes the 'Greatest Transfer of Wealth in History,'" *Foreign Policy*, July 9, 2012, http://thecable.foreignpolicy.com/posts/2012/07 /09/nsa_chief_cybercrime_constitutes_the_greatest_transfer_of_wealth_in_history.

15. For a map of Asia-Pacific cyber censorship, see Morgan Marquis-Boire, Jakub Dalek, Sarah McKune, Matthew Carrieri, Masashi Crete-Nishihata, Ron Deibert, Saad Omar Khan, Helmi Noman, John Scott-Railton, and Greg Wiseman, "Planet Blue Coat: Mapping Global Censorship and Surveillance Tools," Citizen Lab Open Net Initiative, https://citizenlab .org/2013/01/planet-blue-coat-mapping-global-censorship-and-surveillance-tools/.

16. United Nations Office for Disarmament Affairs, "Fact Sheet: Developments in the Field of Information and Telecommunications in the Context of International Security," http://www.un.org/disarmament/HomePage/factsheet/iob/Information_Security_Fact _Sheet.pdf.

17. "Military to Have Initial Cyber Offensive Capabilities in 18 Months, Says Adm. Rogers," *FierceGovernmentIT*, March 5, 2015, http://www.fiercegovernmentit.com/story /military-offensive-capabilities-will-reach-initial-operation-18-months-says/2015-03-05.

18. Doug Drinkwater, "Interpol to Open Cybercrime Centre in Singapore," *SC Magazine*, September 30, 2014, http://www.scmagazineuk.com/interpol-to-open-cybercrime -centre-in-singapore/article/374300/.

19. Sam LaGrone, "Pentagon Drops Air-Sea Battle Name, Concept Lives On," *USNI News*, January 20, 2015, http://news.usni.org/2015/01/20/pentagon-drops-air-sea-battle-name -concept-lives.

20. Richard D. Fisher Jr. and Bill Sweetman, "Sizing Up China's Developing Military Capabilities," *Aviation Week*, April 1, 2011, http://aviationweek.com/awin/sizing-china-s -developing-military-capabilities.

21. Tim Kelly and Nobuhiro Kubo, "Exclusive: Japan Civilian R&D Agency to Get Military Role to Spur Arms Innovation—Sources," Reuters, March 19, 2015, http://www .reuters.com/article/2015/03/19/us-japan-r-d-military-idUSKBN0MF2K520150319.

22. United Nations Office for Disarmament Affairs, "GGE Information Security: Developments in the Field of Information and Telecommunications in the Context of International Security," http://www.un.org/disarmament/topics/informationsecurity/.

23. White House, "Fact Sheet: US–Japan Bilateral Cooperation," April 25, 2014, http:// www.whitehouse.gov/the-press-office/2014/04/25/fact-sheet-us-japan-bilateral -cooperation.

24. Mihoko Matsubara, "Japan–US Cyber Cooperation Needs Urgent Update," *East Asia Forum*, October 22, 2013, http://www.eastasiaforum.org/2013/10/22/japan-us-cyber -cooperation-needs-urgent-update/.

25. "Japan, "United States Includes Cyber Defense in Joint War Exercise," *Recorded Future*, December 9, 2013, https://www.recordedfuture.com/japan-united-states-include -cyber-defense-in-joint-war-exercise/.

26. Russell Brandom, "Cyberattacks Spiked as Russia Annexed Crimea," *The Verge*, May 29, 2014, http://www.theverge.com/2014/5/29/5759138/malware-activity-spiked -as-russia-annexed-crimea.

27. Zachary Keck, "S. Korea Seeks Cyber Weapons to Target North Korea's Nukes," *The Diplomat*, February 21, 2014, http://thediplomat.com/2014/02/s-korea-seeks-cyber -weapons-to-target-north-koreas-nukes/.

28. "Chinese Cyber-Attack on Australia 'Wider Than Previously Thought,'" *The Guardian*, April 28, 2014, http://www.theguardian.com/world/2014/apr/28/chinese-cyber -attack-australia-emails.

29. Australian Ministry of Defence, "Australian Cyber Security Centre to Be Established," January 24, 2013, http://www.defence.gov.au/defencenews/stories/2013/jan /0124.htm.

Part III

Security Solutions

11 Can Multilateralism and Security Communities Bring Security to the Asia-Pacific?

Mathew Davies

Reader's Guide

This chapter introduces both the academic debate about **multilateralism** and the reality of multilateral cooperation in the Asia-Pacific region, distinguishing between different types of multilateral cooperation, including the idea of a **security community**. The chapter first investigates how different theories of international relations explain multilateralism, and it then looks at four key examples of multilateralism: the **Association of Southeast Asian Nations (ASEAN)**, the **ASEAN+3** process, the **ASEAN Regional Forum (ARF)**, and the **East Asia Summit (EAS)**. These examples of multilateralism are characterized by similar shortcomings: a lack of substantive output, the unresolved question of the right size, a lack of leadership, and the role of sovereignty. The chapter closes by considering how increasing US attention to Asia-Pacific multilateralism will lead to both a greater demand for multilateral cooperation and a greater tension about the size, shape, and purpose of this cooperation.

Introduction

The idea of **multilateralism** is a simple one: the coming together of three or more states to address shared political priorities.[1] However, though the idea is easy to understand, the reality of establishing multilateral cooperation in the Asia-Pacific is difficult to achieve. Multilateralism in the region is not as developed as it is in Europe, but there are signs that this is starting to change, especially in Southeast Asia. This chapter explores how multilateralism in the region is shaping relations between the region's states, and discussion reveals that multilateralism is both an opportunity and a challenge. The analysis starts by presenting some of the key ideas about multilateralism and security communities, and then examines key examples of multilateral cooperation in the region: ASEAN, ASEAN+3, the ARF, and the EAS. These discussions serve as the basis for an examination of the strengths and weaknesses of these mechanisms of cooperation. Finally, the discussion turns to examine what the US pivot to Asia means for multilateralism in the region, highlighting how this shift promises to both increase the demand for multilateralism and also make that cooperation more politically contentious than ever.

The Forms and Promise of Multilateralism

Multilateralism can take many forms. It can be as little as a single meeting to discuss a particular issue between a group of states. More usually, however, multilateralism refers to an ongoing series of discussions between states organized around an agreed-on set of aims over an extended period of time. Often, states establish an international organization to facilitate their cooperation, based on legally binding treaties that outline the principles, aims, structure, bureaucratic support (e.g., a secretariat to organize meetings and pass on communication), and some sort of agreed-on diplomatic code that members agree to follow in order to realize their shared aims.[2]

Different theories of international relations view multilateralism in different ways. Realists, believing states are only interested in their own power and security, are skeptical of multilateralism. Military alliances, realists' preferred form of international cooperation, last as long as the power of their major members can support them or as long as the external threat exists. Realists dismiss the ability of multilateralism to bring lasting peace because they believe that this goal is impossible; multilateral cooperation is doomed to be fleeting and precarious, bred of the temporary alignment of diverse states' interests and goals.

Other theoretical approaches are more positive about multilateral cooperation. Liberal accounts suggest that states can be more imaginative in pursuit of their interests than realists believe, and are often particularly interested in the role of economics and trade in driving states to cooperate with each other. Liberals think that multilateralism can help avoid war and promote peace because it brings two things to international politics that are often in short supply: **coordination** and **predictability**. Liberal thought was the driving force behind the creation of many of the international organizations in the twentieth century—the League of Nations in 1919, the United Nations in 1945, and many of the regional organizations created in the Americas, Europe, and indeed Southeast Asia. Multilateralism helps coordinate what states do, developing shared goals and a meaningful way to work toward the realization of these goals. Because of these shared goals, states come to know and understand each other better, and are able to more reliably predict what each other wants, why they want it, and what they will do to achieve it. For liberals, multilateralism helps turn the unknown into the known, which helps decision makers predict with more accuracy what is going to happen, turning enemies into friends and securing peace between them over the long term.

Constructivist accounts offer an even stronger vision of multilateralism that includes the liberal concern with coordination and predictability but also adds the idea of **trust** to the discussion. Trust does not emerge quickly, but is the product of states cooperating over some considerable period of time. Trust is important because it colors how states interpret the behavior of others. Australia, because it trusts the United States as a friend, interprets how the United States moves its military assets around the Pacific in a certain way—a way very different from how China or Russia would interpret exactly the same movements. By building trust, multilateralism can create a **security community**. Karl Deutsch discussed the idea of a security community in the 1950s in relation to the North Atlantic area, arguing that security communities were zones of peace where states had developed reliable expectations of peaceful cooperation in all circumstances and where war thus became unthinkable.[3] Deutsch argued that these security communities came about through processes of "integration," the creation of "unifying habits and institutions" shared by both governments and citizens.[4] These relationships would "lock in" peace by controlling what governments did and creating social bonds of friendship.

Constructivist scholarship built upon Deutsch's argument by emphasizing the role of shared ideas about what states should do and why states should be doing it.[5] Constructivists understand security communities as zones of stable and enduring peace. At their strongest, security communities have eradicated war, or even the threat of war, by making citizens and leaders of countries in that community completely reject war as a policy option between each other.

The Architecture of Asia-Pacific Multilateralism

There are four key multilateral bodies in the Asia Pacific region: ASEAN, ASEAN+3, the ARF, and the EAS. ASEAN is not only the most developed of these organizations but also sits at the heart of the other three (table 11.1).

ASEAN

ASEAN was created in 1967 to promote political and economic cooperation between five states: Indonesia, Malaysia, the Philippines, Singapore, and Thailand. In the early 1960s these states were political and military adversaries; Indonesia and Malaysia were engaged in ongoing border tensions, and the Philippines had claimed parts of Malaysia as its own territory.[6] These were not promising beginnings. But over nearly fifty years, not only has ASEAN expanded to include all the states in Southeast Asia except Timor-Leste; it has also taken on a wider range of cooperation than could ever have been imagined in 1967 (table 11.2).

In 2007 ASEAN members signed the **ASEAN Charter**, a treaty that described what ASEAN was to stand for in the twenty-first century.[7] In Article 1 of this

Table 11.1 Asia-Pacific Multilateral Bodies

Institution	Members	Year Founded
Association of Southeast Asian Nations (ASEAN)	Brunei, Cambodia, Indonesia, Laos, Malaysia, Burma/Myanmar, Singapore, Philippines, Thailand, and Vietnam	1967
ASEAN +3	The above-listed members and China, Japan, and South Korea	1997
East Asia Summit	The above-listed members and Australia, India, New Zealand, Russia, and the United States	2005
ASEAN Regional Forum	The above-listed members and Bangladesh, Canada, the European Union, Mongolia, Pakistan, Papua New Guinea, North Korea, Sri Lanka, and Timor-Leste	1994

Table 11.2 Members of the ASEAN as of 2014

State	Joined	Population	Government	GDP at Purchasing Power Parity (in billions of dollars)
Brunei	1984	417,000	Constitutional sultanate	29.7
Cambodia	1999	15.3 million	Democratic	50.0
Indonesia	1967	254.5 million	Democratic	2,676.2
Laos	1997	6.7 million	Communist	35.6
Malaysia	1967	29.9 million	Democratic	766.6
Burma/Myanmar	1997	53.4 million	Democratizing	N/A
Philippines	1967	99.1 million	Democratic	690.9
Singapore	1967	5.5 million	Democratic	542.7
Thailand	1967	67.7 million	Democratic	1065.7
Vietnam	1995	90.7 million	Communist	510.7

Sources: World Bank, "GDP, PPP (Current International $)," http://data.worldbank.org/indicator/NY.GDP.MKTP.PP.CD; World Bank, "Population, Total," http://data.worldbank.org/indicator/SP.POP.TOTL .

charter, ASEAN's members said there were fifteen things their multilateral co-operation was to facilitate:

- The maintenance of international peace and security;
- Enhancing cooperation in political, security, economic, and sociocul-tural cooperation;[8]
- Preserving Southeast Asia as a zone free of weapons of mass destruction;
- Ensuring a democratic and harmonious environment for the citizens and states of ASEAN;
- Creating a single economic market;
- Poverty alleviation;
- Strengthen democracy, good governance, the rule of law, human rights, and fundamental freedoms;
- Respond effectively to transboundary issues like crime;
- Sustainable development;
- Develop education, science, and technology;
- Enhance the well-being and livelihood of ASEAN citizens;
- Secure ASEAN as a drug-free environment;
- Make ASEAN more people oriented;
- Promote an ASEAN identity; and
- Maintain the centrality and proactivity of ASEAN in regional and extra-regional cooperation.

To help encourage cooperation in these areas, ASEAN now has hundreds of meetings every year, including annual summits between heads of government, regular meetings between national ministers, and a host of technical meetings on particular issues. ASEAN appears to display a sophisticated and successful approach to multilateral cooperation.[9] While it is true that ASEAN now "does more" than ever before, it is also often criticized for its weaknesses. As shown in table 11.2, ASEAN member states are diverse. They range from very small to very large, and some are communist, some democratic, and some democratiz-ing. Singapore is a highly developed, trade-oriented economy with a very high GDP per capita; Thailand, Malaysia, the Philippines, and to a lesser extent Indo-nesia are undergoing massive economic development; and many of the states on the mainland of Asia in Indochina are at comparatively low levels of economic development. These states all have different needs; different relationships with the United States, Japan, and China; and different interests in being part of ASEAN.

This diversity has meant that ASEAN has developed a particular diplomatic environment to regulate both how members relate to each other and how any ASEAN-level activities are run, expressed best in Article 2 of the ASEAN Charter. Most important are commitments to respect "the independence, sovereignty, equality, territorial integrity and national identity" of other members, "non-interference in the internal affairs of ASEAN member-states" and "respect for the right of every member state to lead its national existence free from external interference, subversion and coercion."[10] Alongside these formal rules, ASEAN members also subscribe to social expectations about **diplomacy** that include consensus-based decision making, unanimity, and a preference for informal diplomacy conducted away from the glare of publicity and press attention. The immediate tension between the aims and the diplomatic environment of ASEAN is clear. Can ASEAN successfully help build agreement in areas such as good governance, trade, or poverty alleviation when it is, at the same time, committed to avoiding interfering in the domestic politics of member states?

There is no clear answer to the question of success. ASEAN has two areas of achievement to its name: the maintenance of regional peace and the development of economic cooperation. Since 1967 there has been no major military confrontation in Southeast Asia between member states. This is not to say that tensions have disappeared, but that they have never spilled over into a full-scale war. Similarly, economic cooperation within ASEAN has developed considerably over the decades and now includes substantial trade, resource development, and finance agreements that benefit member states. Notably, ASEAN is in the final stages of realizing a complete free trade agreement between all its members. Yet aside from security and economic cooperation, ASEAN has been far less successful. In the more politically contentious areas listed in the ASEAN Charter, such as human rights, ASEAN's record is far less impressive. The ASEAN Charter has a commitment to human rights promotion, an Intergovernmental Commission on Human Rights was formed in 2009, and an ASEAN Declaration of Human Rights was released in 2012. While this might seem an impressive change, ASEAN members have made sure that the commission and the declaration are compatible with ASEAN's diplomatic environment—most important, that they work with the strong commitment to nonintervention and sovereign equality that has shaped ASEAN since 1967.

If a security community is a zone of peace, and ASEAN has displayed interstate peace since 1967, can it be described as a **security community**? Debate here focuses on what the causes for peaceful relations between ASEAN members might be.[11] Scholars who write supportively of ASEAN suggest that it has encouraged ideas of peaceful coexistence and created a sense of shared regional identity that drives peaceful relations. This work tends toward a constructivist approach, and it identifies norms of cooperation and coexistence that ASEAN has helped create and nurture since 1967. Skeptics are far more critical of ASEAN, and tend toward a realist approach. They argue that relations between ASEAN members are not as peaceful as they seem. There are ongoing border tensions between member states, for example, the Vilhear Temple area between Thailand and Cambodia. Those who study ASEAN military spending note that member states continue to accumulate military resources that are more offensive and targeted against fellow ASEAN members than they would be if there was a security community.[12] As a result of this confusion, some have tried to say that ASEAN, though not a security community, displays some of the characteristics of one and might be developing toward that as an end point.[13]

The ASEAN Regional Forum

The ARF was inaugurated in 1994 in Bangkok, and today is the largest multilateral mechanism—in terms of the number of members—in the Asia-Pacific region. The ARF was created because ASEAN's members in the early 1990s feared that after the end of the Cold War, the United States would decrease its commitment to their security, creating room for Japan and China to compete in the region, and thus threatening the goals of ASEAN unity.[14] To address these concerns, ASEAN's members took their own diplomatic culture that had proven successful in helping their own political tensions (nonintervention, a strong commitment to sovereignty, and a willingness to avoid agreements if even one state disagreed with that position) and sought to widen it to include as many states as possible. The ARF was designed to be a "dialogue-driven process" that would create space for states to talk with each other in a regular fashion without any strong commitments that they would either object to or would create rivalry.[15] These discussions would help enmesh the United States in regional diplomacy, reassuring Japan, South Korea, and many ASEAN member states, while not unduly antagonizing China.

The ARF has two objectives: first, to foster constructive dialogue and consultation between members on political and security issues of common concern; and second, to make a "significant" contribution to efforts toward **confidence building** and **preventive diplomacy**.[16] The ARF is clear that confidence-building measures come before any attempt at preventive diplomacy. Confidence building focuses on starting the process of getting states to view each other with less suspicion because of enduring diplomatic relations within the cooperative framework that the ARF provides. Preventive diplomacy is where states do not wait for a crisis or issue to emerge and threaten peace; rather, they take proactive measures to address issues early. If the ARF worked perfectly, then many of the challenges and issues addressed in this book would either not have arisen or would have been addressed quickly and effectively.

It seems then that the ARF should offer the ideal mechanism for Asia-Pacific multilateralism. However, its track record is very mixed. The simple presence of the ARF might be important, standing as the first step along the path toward trust building by helping to regularize contact between potential rivals and helping create a habit of consultation. However, the ARF has never progressed beyond the first stage of its evolution, confidence building, to deal substantively with the major regional issues via preventive diplomacy. While the ARF meetings talk about the situation on the Korean Peninsula and in the South China Sea, it is hard to see any substantive benefit from these discussions. For example, the chairman's statement regarding the twentieth session of the ARF in July 2013, noted that ministers "underlined the importance of peace, security and stability on the Korean Peninsula" but also noted that only "most" ministers were willing to encourage North Korea to live up to the various United Nations Security Council Resolutions regarding its activities. Indeed, only "most" ministers were even willing to claim that these resolutions needed to be fully implemented at all.[17]

ASEAN+3

ASEAN+3 started with an informal lunch discussion between the ASEAN member states and the "plus three"—China, Japan, and South Korea—in 1994, on the sidelines of the first meeting of the ARF.[18] While ASEAN+3 discusses a full range of political, economic, and social issues, it is the economic field where it has had the most impact. Launched in 2000, the Chiang Mai Initiative has seen ASEAN+3 develop shared financial resources (now some $240 billion) to help combat major currency speculation and the economic problems it caused.[19] The initiative was originally little more than a series of bilateral arrangements where individual countries agreed to help each other. More recently, this has been replaced with a multilateral agreement that all countries contribute money to a central pool, some part of which is then available to those countries if they need it. The Chiang Mai Initiative is credited with helping stabilize the East Asian economy and with facilitating the economic growth the region has experienced.

ASEAN+3 is particularly interesting as it is the only exclusively East Asian multilateral body that brings together both ASEAN members and Northeast Asian states but excludes the United States, other Western states, and India. However, its relationship with other multilateral processes is not clear-cut, and as we shall see when discussing the EAS, different states have different attitudes toward it.

The East Asia Summit

Whereas the ARF is a meeting of the foreign ministers, the East Asia Summit, which met for the first time in December 2005 in Kuala Lumpur, is led by heads of state/government.[20] The idea for the EAS first emerged in a discussion between ASEAN members and China, Japan, and South Korea. At first, China was very keen on the proposal, for it thought it would serve as a mechanism to exclude the United States and cement its own leadership. And though the United States did not join the EAS until 2011 (along with Russia), Japan and some ASEAN members were keen to avoid the EAS becoming dominated by China, and at the outset successfully pushed for the inclusion of India, Australia, and New Zealand.[21]

One of the ideas behind the EAS is that its creation will help to provide new and dynamic leadership. The EAS occurs once a year, immediately after the regular ASEAN Summit of their heads of government. The EAS is open to discussion of a wide range of issues, and has released documents and agreements on climate change, energy, trade, and the environment.

However, the EAS is plagued with considerable problems. It is not clear what its relationship is with other multilateral bodies, especially the ASEAN+3 system. China in particular is keen to emphasize ASEAN +3 over the EAS, as it promises a greater opportunity for Beijing to develop leadership excluding both Western powers and India. Others, notably the United States but also Australia, are keen that the EAS should exercise overall leadership because they are a part of that system and fear being excluded. Japan and South Korea, though members of both, are wary of putting too much emphasis on ASEAN+3 in case that facilitates China's leadership to their own detriment.

These forums are known as **track-one** diplomacy, involving the formal meetings of state representatives. However, these are far from the only meetings that occur. **Track-two** diplomacy, involving the activities of nonstate actors such as individuals and organizations, also plays an important role in the Asia-Pacific region because of the close, if informal, relationship between it and formal interstate negotiations.[22] For example, it was a track-two body, the Council for Security Cooperation in the Asia-Pacific, that first developed a definition of "preventive diplomacy" for the ARF. In ASEAN a group called ASEAN-ISIS (Institutes of Strategic and International Studies) has played a similar role. Being made up of leading universities and think tanks in ASEAN member states, ASEAN-ISIS provides new insights into regional issues from an informal but influential perspective. Track-two diplomacy often includes many of the same men and women who work in track one, who attend meetings in an "informal" or "personal" capacity. This close relationship helps ideas that otherwise might be too contentious to emerge in track-one discussions be debated and agreed upon. Track-two diplomacy is evident in security, economic, and social areas of regional cooperation.[23]

Challenges

There are four challenges common to ASEAN, ASEAN+3, the ARF, and EAS: (1) that they are all talk but no action; (2) tensions over the membership of the organizations, particularly whether they should be East Asian or Asia-Pacific; (3) how to balance sovereignty and effectiveness; and (4) the question of leadership.

All talk but no action? One of the criticisms leveled at all the multilateral forums discussed in this chapter is that they do little more than provide "talking shops" where lots of words are spoken but very few, if any, substantive decisions are made. Would East Asia look any different if there was no ASEAN, ASEAN+3, ARF, or EAS? There would certainly be far less diplomatic contact between the states in the area, but would their relationship be fundamentally worse than it is today? Put another way, what is the actual value of Asia-Pacific multilateralism?

ASEAN's engagement with human rights is an excellent example of this weakness. That ten diverse member states were able to agree on human rights issues and place them into the regional body is evidence that ASEAN's multilateral approach can achieve agreement in even the most contentious areas. However, many wonder whether these agreements will result in any actual change. The commitment to nonintervention means that domestic politics, where the vast majority of human rights abuses occur, is off limits to the commission. The commitment to nonintervention has meant that countries with communist or authoritarian governments have been happy to agree with human rights reforms because they know that their domestic political activities are off limits, and so any regional agreement is highly unlikely to ever be enforced.[24]

The answer to whether multilateralism achieves anything depends on how much you value the processes of talking that lead up to these agreements.[25] Critics argue that the talk that happens in these organizations is pointless, a lot of hot air with no substantive outcome. Even where there are agreements on paper, these mean little and are more for show than because of an honest commitment.[26] More-optimistic analysts argue that the "process-oriented" approach to multilateralism in the Asia-Pacific is a unique solution to a unique set of problems. Yes, there is a lot of talk without much agreement, but this talk is in itself very important. The highly diverse membership of ASEAN, let alone the other multilateral examples just discussed, requires a light touch and slow approach. The endless talking that the critics think such a waste of time is in fact the whole point of these types of cooperation, helping states that otherwise would be suspicious of each other to come together and discuss various topics with each other. Yes, the supporters admit, there is not a lot of output from some of these bodies, but to take that fact and say they were irrelevant would be to misunderstand what they bring to international politics. Talking promotes regular contact and helps build a habit of multilateralism and can be the crucial first step toward **trust-building** activities and preventive diplomacy.

East Asian or Asia-Pacific? Countries in the region disagree about the ideal shape of multilateral cooperation. Should there be a single multilateral body of which all states are members, or should there be a series of smaller ones? While different multilateral options in the Asia-Pacific have different patterns of membership, no one organization has a pattern that makes all participants equally happy. This has created significant dissatisfaction on the part of many states with the extent of multilateral cooperation in the region, but little consensus on the way to address that.

These questions touch on one of the paradoxes of multilateralism. By including more actors in more roles, and by discussing more issues, multilateralism can make it very hard to generate any substantive agreements.[27] We might prefer multilateralism to be as inclusive as possible, because that would mean that any agreements reached would cover a large area. The history of the ARF and EAS suggests that most states in the region are also interested in being on the inside of any multilateral system rather than being excluded from it. However, the very act of including more states, especially in an area as diverse as the Asia-Pacific, makes that agreement more difficult to achieve. Exclusive multilateralism, including only a few states, may offer the opportunity to develop more meaningful and substantive cooperation. It is not a coincidence that the two smaller multilateral bodies discussed, ASEAN and ASEAN+3, have produced far more cooperation than the more inclusive ARF or EAS.

Sovereignty versus effectiveness. One of the problems that emerges when there are multilateral bodies that have such a variety of members is that the approach of these organizations becomes more limited. In ASEAN, ASEAN+3, the ARF, and EAS, ideas about sovereignty and nonintervention have come to the fore. These ideas dominate because they are the easiest way to reassure states that they will not be criticized at meetings or be forced to agree to things to which they do not want to agree. By taking domestic affairs off the discussion list, states that otherwise disagree over a great many things can be brought together.

However, if domestic affairs are not to be discussed, then a great number of the causes of disagreement and dispute in the Asia-Pacific region are also left unaddressed. Most notably, Taiwan is excluded as a member and topic of conversation from all four multilateral bodies because China views it as not a separate state but as a renegade province and thus not as an international issue but a domestic one. More generally, we have seen how ASEAN and the ARF have moved to tackle "nontraditional" security concerns, such as the environment, poverty, and, in ASEAN's case, even human rights and democracy. Here the norms of nonintervention that shape Asia-Pacific multilateralism seem to actively work against the successful realizing of some of the goals that multilateralism in Asia now sets itself. How can one protect or promote human rights without talking about the domestic politics of states, given that the state itself is most often the cause of human rights abuses?

The question of leadership. The question of leadership is a vexed one. A clear, and agreed-upon, leading state would help generate momentum and direction for any of the Asia-Pacific multilateral bodies discussed, both of which are lacking in the current system. However, which state that could or should be is highly contentious. Washington insiders might wish it were the United States, but this is unlikely to generate any agreement in Beijing, and even pro–United States ASEAN states do not want to see the United States replace their own current leadership of either the ARF or EAS. The relative decline of the United States, especially when gauged against China, suggests that Washington cannot now simply force the issue, as it was able to in post–World War II Europe. And the most logical alternative, Beijing, is viewed with considerable distrust by other states in Northeast and Southeast Asia. There seems little chance that Tokyo or Jakarta would acquiesce to untrammeled Chinese leadership. This tension helps explain why the ASEAN member states maintain their leadership in Asia-Pacific multilateralism. ASEAN leadership, such as it is, is not an agreed-upon decision, reached after careful consideration. Instead it is the default option, given that no agreement on an alternative can be reached. Placing smaller, less powerful states in charge has exacerbated the sense of drift that characterizes much regional diplomacy. Because they are too small to enforce their wishes, ASEAN members are forced to adopt the slowest, most open-ended, and least confrontational approach, undercutting the ability of multilateral cooperation to meaningfully address common concerns.

Using Theories to Understand Asia-Pacific Multilateralism

Realist, liberal, and constructivist analyses are all valuable for understanding the nature of Asia-Pacific multilateralism, and instead of trying to argue which one theory is right, we should ask how each of the theories highlights different aspects of what is happening. Theories are tools that we can use to help bring different parts of a complex world into focus.

Realist scholarship helps us identify the importance of national interest in all these forms of cooperation, and reminds us to be skeptical of the agreements that are so quickly made and so rarely actually realized. Realist scholarship is well positioned to explain some of the weaknesses discussed above—the selfish preoccupations of the major states in the region, the limited ability of these organizations to bring about peaceful cooperation, and the apparent failure of so many of the ventures upon which these organizations embark.

Yet liberals, too, bring something to the discussion. The lasting peace in Southeast Asia, driven in part by economic cooperation, is hard for realists to explain. Even if one might doubt the effectiveness of many of the organizations discussed, their very presence and longevity seem to suggest that there is something more going on than simple realist concerns with power. Liberals would also suggest that the presence of all these organizations, even if they fail so often, reveals an important truth: that there is a desire for some sort of international cooperation that realists will never understand. Organizations are the logical response for states in the region as they deal with the complex issues they face.

Finally, though the constructivist vision of shared norms and lasting peace seems too optimistic to describe the situation in the Asia-Pacific at the moment, it does offer insight into some of the actions and motives of states, especially in Southeast Asia. Efforts to embed norms of cooperation, consensus, and the use of diplomacy are about creating common understandings that build trust. Constructivists provide reasons for optimism despite the current situation because they see how states can come together in new ways.

The US Pivot and Asia-Pacific Multilateralism

In January 2010 then–US secretary of state Hillary Clinton spoke on "regional architecture in Asia" at the East-West Center in Honolulu. Clinton claimed not only that the United States saw its future in the Asia-Pacific region but also that it had a "strong interest in continuing its tradition of economic and strategic leadership."[28] The US pivot toward the region rests on three dimensions. First, the pivot represents a reallocation of military resources to the Pacific from other theaters to underline US power. Second, the United States has sought to reclaim diplomatic leadership by recommitting to the EAS and investing both money and personnel in developing a stronger presence in many regional capitals.[29] And third, the pivot is a claim to economic leadership, especially via new trade agreements. The United States' intention is to reinvigorate Asia-Pacific multilateralism to help facilitate the US strategic interest in keeping the region both peaceful and conducive to US interests.

Perhaps the most significant achievement in this direction is the resolution of the Trans-Pacific Partnership (TPP) negotiations in October 2015. The TPP is not a formal multilateral institution, but a trade agreement that commits twelve Pacific Rim countries to similar standards for interstate trade. It will necessitate closer dialogue and regular interactions between the countries involved. The members are Australia, Canada, Japan, Malaysia, Mexico, Peru, the United States, Vietnam, Chile, Brunei, Singapore, and New Zealand. Supporters of the deal believe the TPP will increase the integration of regional economies with the United States, building resilience into the political alignment of these countries with American leadership. Opponents worry about the distortion of trade, and whether the TPP might encourage rival trading blocs, given that China, Indonesia, and other prominent economies were not involved in the initial agreement.

Under the aegis of the EAS, US leadership would promote the resolution of long-standing political and security concerns. This would not only serve the interests of Washington but also those of key regional states, defusing the tensions that plague their mutual relations and promoting economic cooperation. However, the idea of US leadership through the pivot is uncomfortable not just to China but also to others. Beijing has little interest in ceding leadership to the United States; hence its particular interest in ASEAN+3 over the EAS. As discussed above, the ASEAN states make it a key aim that they maintain leadership in all the Asia-Pacific multilateral systems discussed in this chapter. Efforts by the United States to replace ASEAN's centrality would not only be resisted but would also lead to considerable political tensions.

The most dangerous consequence of the pivot is that it falls between two chairs—marking the strengthening of the US presence in the region but also the failure of Washington to claim and maintain decisive leadership. Such an outcome would antagonize China by curtailing its own bid for leadership while not forcing China to cooperate within any revised system. Given China's importance, and the intergovernmental workings of all Asia-Pacific multilateralism, China would at a minimum be able to refuse to cooperate, stopping any agreements it felt were detrimental to its own interests. At worse, regional states would be forced to choose sides between China and the United States. This would be a disaster for the current multilateral system, which is based on states coming together and not choosing sides. Australia, for example, claims that there is no tension between its ongoing security alliance with the United States and its ever-deeper economic relationship with China. An incomplete pivot, if it were to back-fire and cause rather than solve tensions, would force Canberra, and many other states in similar situations, to choose.

Conclusion

The states of the Asia-Pacific are a long way from realizing enduring and meaningful multilateralism across the entire region. Only in Southeast Asia is there sufficient evidence to say that multilateralism developed and helped achieve common goals; but even ASEAN is often criticized. The discussion in this chapter has highlighted the key weaknesses of the various efforts to manage regional affairs. The enduring tensions in the Asia-Pacific, the rivalry between key players, and their seeming disinterest in shaping effective multilateral organizations all stand as formidable, perhaps even insurmountable, obstacles.

Under these circumstances, one must ask whether the region's existing multilateral organizations—ASEAN, ASEAN+3, the ARF, and the EAS—are adequate to the task of ensuring even a fragile and temporary peace between the key actors. The track record of the organizations discussed here is mixed; optimists point to the absence of major conflict, whereas pessimists point to the enduring tensions of the region. Looking to the future, the US pivot to the region seems highly unlikely to address any of the shortcomings in the region's multilateralism. The pivot threatens to exacerbate the tensions and disagreements that currently hamstring multilateral cooperation by raising the stakes for all parties. If multilateralism is to succeed in the region, a renewed political commitment to diplomacy and not the rhetoric and practice of nationalism will be required. Tough questions will need to be asked about the current system and what revisions to membership, leadership, and decision making procedures may be required. Whether states in the region are willing to overhaul both their own national priorities and the diplomatic behavior they use to pursue these aims will be a key question in the early twenty-first century. The answer to this question will reveal whether the promise of multilateralism can be fulfilled.

Key Points

- Multilateralism in the Asia-Pacific is very unevenly distributed. ASEAN is by far the most developed example, in terms of both the range of issues that are covered and the frequency of these discussions.

- All Asia-Pacific multilateral bodies have been critiqued as producing too much talk but not enough substantive output.
- There is a lack of multilateralism explicitly designed by, and for the needs of, the great powers of the region. There is no organized multilateral organization that contains only the most important states—China, South Korea, Japan, and the United States.
- This absence of great power multilateralism in the region worries many people because it limits the ability of key states to directly negotiate about their problems. This spreads uncertainty, unpredictability, and distrust over the motives, intentions, and interests throughout the system.

Questions

1. Is ASEAN an example of successful multilateralism? Is it a security community?

2. Are ASEAN+3, the ARF, and EAS adequate to promote peaceful relations in the region?

3. Can ASEAN members continue to lead in Asia-Pacific organizations? Is their leadership a positive development, or does it show that the great powers do not care?

4. Will the American pivot to the Asia-Pacific exacerbate political tensions between states to the point that multilateral efforts break down? Or will it actually reinvigorate these organizations?

5. Why are there no general multilateral organizations designed for and by the great powers of the region?

6. Would an organization comprising just China, Japan, North Korea, South Korea, Russia, and the United States be better at solving regional tensions than the current system?

7. How can the opportunities for multilateralism be expanded in the Asia-Pacific?

Guide to Further Reading

Acharya, Amitav. *Constructing a Security Community in Southeast Asia: ASEAN and the Problem of Regional Order*. London: Routledge, 2001. http://dx.doi.org/10.4324 /9780203393345.
This classic book explores the history of Southeast Asian multilateralism and tests to what extent ASEAN can be understood as a security community.

Ba, Alice D. *(Re)negotiating East and Southeast Asia: Region, Regionalism and the Association of Southeast Asian Nations*. Stanford, CA: Stanford University Press, 2009.
An excellent and detailed analysis of the history and evolution of ASEAN and East Asian multilateralism. Ba takes the opportunity to "test" East Asian multilateralism against key theories of international relations.

Katsumata, Hiro. "The Establishment of the ASEAN Regional Forum: Constructing a 'Talking Shop' or a 'Norm Brewery'?" *Pacific Review* 19, no. 1 (2006).
An excellent overview of the rival approaches to just how important talking without a substantial outcome might be. Katsumata takes issue with the skeptical realist account and suggests that a constructivist approach is better for understanding the true value of both open-ended debate and the ARF.

Kerr, Pauline, and Geoffrey Wiseman. *Diplomacy in a Globalizing World: Theories and Practice.* Oxford: Oxford University Press, 2013.
A recent textbook that explores the past, present, and future of diplomacy with a particular Asia-Pacific focus. Includes thematic chapters on security and on economic and track-two diplomacy alongside country studies of China and the United States.

Ravenhill, John. "East Asian Regionalism: Much Ado about Nothing?" *Review of International Studies* 35, no. 1 (2009): 215–35. http://dx.doi.org/10.1017/S0260210509008493.
A very good overview of the debate over whether East Asian multilateral cooperation produces significant outcomes. The article focuses on the academic debate and helpfully reveals the key patterns of agreement and disagreement.

Roberts, Christopher B. *ASEAN Regionalism: Cooperation, Values, and Institutionalization.* London: Routledge, 2012.
Roberts takes a long view of the history and evolution of ASEAN and provides a critical investigation into how and why ASEAN has taken the form that it has. There is some useful analysis of different ideas of regional order and cooperation to help frame these discussions.

Notes

1. Robert O. Keohane, "Multilateralism: An Agenda for Research," *International Journal* 45 (1989): 731.

2. Richard Higgott, "International Political Institutions," in *The Oxford Handbook of Political Institutions*, ed. R. A. W. Rhodes, Sarah A. Binder, and Bert A. Rockman (Oxford: Oxford University Press, 2008), 614.

3. Karl W. Deutsch, *Political Community at the International Level* (Garden City, NY: Doubleday, 1954), 33.

4. Karl Deutsch, *The Analysis of International Relations, Second Edition* (Englewood Cliffs, NJ: Prentice Hall, 1978), 194.

5. Emanuel Adler, "Imagined (Security) Communities: Cognitive Regions in International Relations," *Millennium-Journal of International Studies* 26 (1997): 249–77; Emanuel Adler and Michael Barnett, *Security Communities* (Cambridge: Cambridge University Press, 1998).

6. Nicholas Tarling, *Regionalism in Southeast Asia: To Foster the Political Will* (London: Routledge, 2006), 97–140.

7. ASEAN, "The ASEAN Charter," http://www.asean.org/storage/images/ASEAN_RTK _2014/ASEAN_Charter.pdf.

8. ASEAN's website provides an excellent overview of ASEAN, its evolution, and its current activities; see http://www.asean.org/asean/about-asean/overview/.

9. Mely Caballero-Anthony, "Re-Visioning Human Security in Southeast Asia," *Asian Perspective* 28 (2004): 155–89.

10. ASEAN, "ASEAN Charter."

11. John Ravenhill, "East Asian Regionalism: Much Ado about Nothing?" *Review of International Studies* 35, no. S1 (2009): 215–35.

12. Y. U. Wang, "Determinants of Southeast Asian Military Spending in the Post–Cold War Era: A Dynamic Panel Analysis," *Defense and Peace Economics* 24, no. 1 (2012): 73–87; Robert Hartfiel and Brian L. Job, "Raising the Risks of War: Defense Spending Trends and Competitive Arms Processes in East Asia," *Pacific Review* 20, no. 1 (2007): 1–22.

13. Amitav Acharya, *Constructing a Security Community in Southeast Asia: ASEAN and the Problem of Regional Order* (London: Routledge, 2001).

14. Alice Ba, *(Re)negotiating East and Southeast Asia: Region, Regionalism, and the Association of Southeast Asian Nations* (Stanford, CA: Stanford University Press, 2009), 160.

15. Ibid., 180.

16. ASEAN Regional Forum, "About Us," http://aseanregionalforum.asean.org/about.html.

17. ASEAN Regional Forum, "Chairman's Statement of the 20th ASEAN Regional Forum," http://www.asean.org/storage/images/Statement/chairmans%20statement%20of%20the%2020th%20asean%20regional%20forum%20-%20final.pdf.

18. A record of ASEAN+3 cooperation is available at http://www.asean.org/asean/external-relations/asean-3.

19. Ba, *(Re)negotiating*, 228–30.

20. A record of the EAS is available at http://www.asean.org/asean/external-relations/east-asia-summit-eas/.

21. Mohan Malik, "The East Asia Summit," *Australian Journal of International Affairs* 60 (2006): 209–10.

22. Pauline Kerr and Brendan Taylor, "Track-Two Dipomacy in East Asia," in *Diplomacy in a Globalizing World: Theories and Practice*, ed. Pauline Kerr and Geoffrey Wiseman (Oxford: Oxford University Press, 2013), 226.

23. Desmond Ball, Anthony Milner, and Brendan Taylor, "Track 2 Security Dialogue in the Asia Pacific: Reflections and Future Directions," *Asian Security* 2, no. 3 (2006): 174–88; Mathew Davies, "Explaining the Vientiane Action Programme: ASEAN and the Institutionalisation of Human Rights," *Pacific Review* 26, no. 3 (2013): 385–406.

24. Mathew Davies, "ASEAN and Human Rights Norms: Constructivism, Rational Choice and the Action-Identity Gap," *International Relations of the Asia Pacific* 13, no. 2 (2013): 207–31; Mathew Davies, *Realising Rights: How Regional Organizations Socialise Human Rights Norms* (London: Routledge, 2014); Mathew Davies, "States of Compliance? ASEAN and Global Human Rights Treaties," *Journal of Human Rights* 13, no. 4 (2014): 414–33.

25. Hiro Katsumata, "Establishment of the ASEAN Regional Forum: Constructing a 'Talking Shop' or a 'Norm Brewery'?" *Pacific Review* 19 (2006): 181–82.

26. David Martin Jones and Michael L. R. Smith, "Making Process, Not Progress," *International Security* 32 (2007): 148–84; Nicholas Khoo, "Deconstructing the ASEAN Security Community: A Review Essay," *International Relations of the Asia Pacific* 4 (2004): 35–46.

27. I. William Zartman, "Diplomacy as Negotiation and Mediation," in *Diplomacy in a Globalizing World: Theories and Practice*, ed. Pauline Kerr and Geoffrey Wiseman (Oxford: Oxford University Press, 2013), 114.

28. Hillary Clinton, "Remarks on Regional Architecture in Asia: Principles and Priorities," speech, Honolulu, January 12, 2010.

29. Wei Ling, "Rebalancing or De-Balancing: US Pivot and East Asian Order," *American Foreign Policy Interests* 35 (2013): 150.

12 Is Human Security a Solution?

Sarah Teitt

Reader's Guide

This chapter examines whether embracing a more people-centered approach to security in the Asia-Pacific—that shifts the focus from defending sovereign borders to safeguarding the region's populations—offers a solution for achieving security in the region. It explores how traditional state-centric security, which focuses on the military defense of sovereign territory, fails to address some of the most pressing security challenges that the region's states and peoples face. With these challenges in mind, the chapter examines how new people-centered security norms and practices that seek to advance human development and prioritize human protection can help address some of the threats posed by rapid economic development, natural disasters, and armed conflicts in the region. The chapter points to a need to adopt a more comprehensive, people-centered approach to security that safeguards the region's most vulnerable populations so that the path to securing the region is stable, sustainable, and just.

Introduction

During the last two decades, the concept of human security has served to amend and expand the field of security studies. Whereas traditional security studies focuses on the role of military arms in defending sovereign territory, human security is concerned with safeguarding individual human beings from vulnerabilities arising from extreme poverty and armed conflict. Human security shifts the basic question from "how do we defend states from military attack?" to "how do we ensure that everyday human beings live free from misery and massive violence?" This shift exposes for scrutiny how upholding fundamental human rights and meeting basic human needs can help secure peoples and states. Human security spans economic, social, and political policy areas, and during the past decade it has focused on advancing progress on the **Millennium Development Goals** (eight

Table 12.1 The Millennium Development Goals

Number	Goal
1	Eradicate extreme poverty and hunger
2	Achieve universal primary education
3	Promote gender equality and empower women
4	Reduce child mortality
5	Improve maternal health
6	Combat HIV/AIDS, malaria, and other diseases
7	Ensure environmental sustainability
8	Global partnership for development

international development goals agreed upon by the international community at the United Nations Millennium Summit in 2000), the **Sustainable Development Goals** (which build on and expand the Millennium Development Goals and were agreed upon by the international community in 2015; see table 12.1), and implementing the **Responsibility to Protect (R2P)** principle (a commitment to take action to protect populations from genocide, war crimes, ethnic cleansing, and crimes against humanity, which the UN member states unanimously endorsed at the 2005 UN World Summit). The Millennium Development Goals, Sustainable Development Goals, and R2P principle have been influenced by the major international human rights conventions, which enumerate the international legal foundations of human security (see table 12.2).

The Asia-Pacific region has long embraced a relatively holistic view of security that looks beyond territorial defense from military aggression, to include economic and sociopolitical issues. However, comprehensive security concepts that originate in the region tend to place paramount importance on the preservation and promotion of the security of the state as the ultimate social "good," and to emphasize the primacy of sovereignty and noninterference in interstate relations. The region has therefore proven slow to adopt a human security approach as a means for achieving security, which would require focusing on people rather than states as the primary referent object of security, and accepting that the right to sovereign noninterference is qualified by the responsibility to safeguard domestic populations from extreme harm.

This chapter examines whether embracing a more people-centered approach to security in the Asia-Pacific region—shifting the focus from defending sovereign borders to safeguarding the region's populations—offers a solution for achieving security. It begins with an overview of the human security concept, and then examines how human security resonates in the Asia-Pacific security context. The chapter then outlines how more people-centered security policies could help address the security threats imbedded in, and arising from, the region's

Table 12.2 The Major International Human Rights Conventions

Name	Date	Asia-Pacific Signatories
Genocide Convention	December 9, 1948	All except Bhutan, Brunei, Indonesia, Japan, Marshall Islands, Federated States of Micronesia, Palau, Samoa, Solomon Islands, Tajikistan, Thailand, Turkmenistan, and Vanuatu
Universal Declaration of Human Rights	December 10, 1948	Adopted by United Nations General Assembly
Refugee Convention	July 28, 1951	Afghanistan, Australia, Cambodia, Canada, China, Fiji, Japan, Kazakhstan, Kyrgyzstan, Nauru, New Zealand, Papua New Guinea, the Philippines, South Korea, Russia, Samoa, Solomon Islands, Tajikistan, Timor-Leste, Turkmenistan, and Tuvalu
International Covenant on Civil and Political Rights	December 16, 1966	All except Bhutan, Brunei, China, Fiji, Kiribati, Malaysia, Marshall Islands, Federated States of Micronesia, Myanmar, Nauru, Palau, Singapore, Solomon Islands, Tonga, and Tuvalu
International Covenant on Economic, Social, and Cultural Rights	December 16, 1966	All except Brunei, Bhutan, Fiji, Kiribati, Malaysia, Marshall Islands, Federated States of Micronesia, Nauru, Palau, Samoa, Singapore, Tonga, Tuvalu, the United States, and Vanuatu
Convention on the Elimination of all Forms of Discrimination Against Women	December 18, 1979	All except Palau, Tonga, and the United States
Convention on the Rights of the Child	November 20, 1989	All except the United States

economic development model. The final section examines the threats posed by armed conflict and humanitarian disasters in the region, and provides insight into how the R2P principle may help secure Asia-Pacific populations by building a regional commitment and capacity to prevent mass atrocities. The chapter concludes that prioritizing human development and taking strides to implement the R2P principle are important steps to ensuring that the path to securing the Asia-Pacific is stable, durable, and just.

The Concept of Human Security

The concept of human security was introduced to international security discourse in the 1994 *Human Development Report* issued by the United Nations Development Program (UNDP). The UNDP report argued that the traditional concept of security, which was preoccupied with protecting sovereign territories from external military threats, was outmoded in the era of globalization and failed to account for some of the most pernicious threats to the lives and livelihoods of the majority of the world's population. The UNDP called for a new focus on "human security," which was a term meant to focus attention to the severe insecurities that ordinary people encounter in their everyday lives due to extreme poverty and armed conflict. Human security emerged as a concept that encapsulated policies aimed at protecting populations from the effects of armed conflict, addressing basic human needs, and assisting impoverished and marginalized populations with social safety nets. The UNDP's definition of human security expanded the scope of security in seven component policy areas: economics, food, health, the environment, personal, community, and political security. Human security thereby spanned the economic, social, and political policy areas, and privileged individuals and communities, rather than states, as the primary referent object of security. The UNDP's human security agenda implied that in addition to deterring military threats to states, security policy should be geared to protect populations within states.

There are five key aspects that separate human security from the traditional state-centric, military-focused definition of security. First, human security is *people-centered*; it holds that the definition of security should apply to individuals and communities and should be aimed at the protection and empowerment of populations. Second, it is *universal*; human security rests on universal life claims that all human beings are the bearers of the basic right to live in dignity. Third, it is *multivocal*; it recognizes that security actors include a wide range of state, non-state, and civil society agents. Fourth, it is *multidimensional*; it recognizes that security has implications for a range of economic, political, and social policies areas. And fifth, it is *interdependent*; as human security threats transcend state borders, addressing them is a collective responsibility that requires transnational solidarity and cooperation. Together, these aspects point to the overarching objective of human security: to ensure the survival and dignity of individual human beings. This objective serves to diminish the importance of international borders and implies that security agents and institutions bear responsibilities toward both domestic and international populations.

As the concept of human security gained a foothold in security discourse during the 1990s, debates arose over the focus of the human security policy agenda. The UNDP's report defined two strands of human security: freedom from fear ("safety from chronic threats such as hunger, disease, and repression") and freedom from want ("protection from sudden and hurtful disruptions in the patterns of daily life—whether in homes, in jobs, or in communities").[1] The governments of Canada and Japan respectively became key state advocates of each of these strands of human security, with Canada calling for a narrower focus on protecting populations from violent conflict and war, and Japan advocating for a broader poverty reduction and human empowerment agenda.

Canada sought a narrow "freedom from fear" focus in which human security applied mainly to safeguarding individuals and communities from systematic and endemic harm caused by interstate war, state-sponsored violence, genocide, crimes against humanity, and ethnic or communal conflict. Specific issues under this approach to human security focus on the protection of civilians from violent conflict and war, and include such efforts as providing emergency humanitarian relief and assistance to conflict-affected populations; banning indiscriminate weapons, such as land mines; prosecuting perpetrators of large-scale human rights abuses; advancing women's peace and security; deploying international peacekeeping operations; intervening to halt genocidal violence; and improving conflict prevention and peace-building initiatives. As part of its human security advocacy to protect civilians from mass atrocities, as witnessed in Rwanda in 1994 and Kosovo in 1999, Canada sought to build a consensus and the political will for "**humanitarian intervention**" (which is defined as coercive military action taken against a state for the purposes of protecting people from grave human rights violations being perpetrated by that state). Because of its focus on contravening sovereignty to uphold international human rights and international humanitarian law, under Canada's human security advocacy, "freedom from fear" became associated with the "Western" liberal human rights agenda that prioritizes the protection of individual civil and political rights, especially during armed conflict.

In contrast to Canada's "freedom from fear" human security advocacy, Japan focused on the broader or deeper "freedom from want" agenda. Japan's advocacy was motivated by the confronting fact that more people die on a daily basis as a result of extreme poverty and underdevelopment than from conflict-related violence. On this basis, Japan's approach to human security called for greater attention to the myriad economic, food, health, environmental, and community insecurities that render people's everyday lives miserable and insecure. It focused on those issues of structural violence associated with severe hunger, a lack of shelter, disease, a lack of education, social discrimination, and inequality. Whereas Canada's advocacy was viewed as a means to advance Western liberal individual human rights, Japan's human security advocacy became associated with the Global South's communal human rights agenda, which prioritizes the promotion of social and economic rights.[2]

While there appears to be significant divergence in what have become known as the "Canadian" versus "Japanese" strands of human security, they share the same central premise: that security is bereft of meaning if it does not address egregious human suffering. Central to both concepts of human security is the shift from the defense of a territorially bounded state to the protection and empowerment of human beings. For both strands, human security is an integrative rather than a defensive concept, and material or physical threats to the state are intertwined with broader threats to human life and human dignity.

Human Security in the Asia-Pacific

How does human security resonate with the Asia-Pacific security context? Although the 1994 UNDP report claimed to introduce a "new" security concept, expansive notions of security were not novel in the Asia-Pacific security lexicon. As described in chapter 3, in the early 1980s Japan coined the concept of **comprehensive security**, which looked beyond the narrow focus on the self-defense of independent states to also include food, energy, environmental, and social security threats as regional and global security challenges. Earlier still, in the 1960s and 1970s—the height of the Cold War—Southeast Asian states such as Indonesia, Malaysia, and Singapore endorsed multidimensional security concepts that encompassed a wide range of development and security challenges to post-independence/postcolonial states, including economic recession, secessionist and insurgency movements, and communal and ethnic conflict.[3] Indeed, the Asia-Pacific region has a relatively enduring history of embracing a more holistic view of security that looks beyond territorial defense from military aggression, and spans both domestic and foreign policy arenas.

Despite a relative comfort with broader definitions of security, human security has encountered a number of normative and practical roadblocks in the Asia-Pacific. Resistance is in large part due to the region's preoccupation with national security and state-led development. Many states in the region adopt a form of state developmentalism, which places rapid economic growth (rather than human empowerment and protection) at the center of political activity and justifies soft authoritarian rule and a strong, centralized government.[4] Rather than recognizing individuals as the referent object of security, broader definitions of security that originate in the region focus almost exclusively on the preservation and promotion of the security of the state as the ultimate social good.[5] Although less prominent than in the early 1990s, regional leaders have also promoted an "Asian values" debate on human rights, which prioritizes communal rights over individual rights, and defines "community" as the sovereign state. Asia-Pacific leaders have therefore been generally impervious to human protection norms generated outside the region, and have viewed the "freedom from fear" human security agenda as "a Western agenda, centering on such liberal values and approaches as human rights and humanitarian intervention [that give] short shrift to the economic and development priorities of the region."[6] Dewi Fortuna Anwar characterizes the region's security paradigm as a state-centered and community-oriented approach that is "predominately inward looking," which "tends to relegate human rights and human security to secondary importance" and allows for the suppression of individual rights in the name of state security.[7]

Despite skepticism about the liberal individualism undergirding human security, the Asia-Pacific states have recognized that state stability and people's security are intertwined. Amitav Acharya argues, for example, that the devastating effects of economic crises, communicable diseases, and natural disasters have underscored the need to adopt a more people-centered approach to security. Three events stand out in contributing to changes in the region's security mindset. The first is the 1997 Asian financial crisis, which was marked by a major depreciation in the value of currencies in the Asia-Pacific region, and revealed severe problems with the unregulated practices of banking and financial sectors and the high level of foreign debt that had fueled impressive economic growth rates in many Asian countries in the 1990s. The 1997 Asian financial crisis witnessed widespread riots in Indonesia over the drastic devaluation of the rupiah and corresponding sharp increases in consumer prices. Popular unrest in Indonesia brought on by the economic crisis ultimately spurred the collapse of the Suharto regime, which had ruled Indonesia for three decades. The experience in Indonesia reinforced a perception among Asia-Pacific leaders that "the security of authoritarian developmental regimes cannot be insulated from crises that produce large-scale human suffering."[8] The second event that underscored the need to adopt more people-centered security practices in the Asia-Pacific was the 2002 outbreak of Severe Acute Respiratory Syndrome (SARS), a contagious and sometimes fatal respiratory illness that originated in southern China and spread to other countries, and resulted in around 8,273 infected people and some 775 deaths.

The Chinese government attracted widespread international criticism for withholding information about the SARS outbreak and for failing to contain the spread of the illness. International commentators called into question the ability of China's secretive, top-down governing regime to openly manage health crises and prevent a transnational contagion. For leaders in Beijing and throughout the region, the SARS crisis underscored that a nontransparent, insular, and underprepared public health system could lead to international political fallout that bore implications for domestic stability. The third event that compelled the Asia-Pacific states to embrace more people-oriented security policies in the region was the December 2004 Indian Ocean tsunami, in which a major earthquake led to tsunamis that inundated coastal communities along the Indian Ocean. This tsunami resulted in nearly 230,000 deaths in fourteen countries, which amounted to one of the world's deadliest recorded natural disasters. A massive international humanitarian aid effort was launched to assist the worst-hit communities in Indonesia, Sri Lanka, India, and Thailand. The scale of the devastation and high death toll reinforced the need to consider natural disaster preparedness as a key matter of regional security cooperation. In sum, the scope and severity of each of these crises, and the economic and political stressors they placed on the states involved, had the effect of deepening the conviction among leaders in the Asia-Pacific that state security and human security are mutually reinforcing rather than antagonistic, because grave human insecurity "puts the state in permanent danger of disintegration and collapse."[9]

Although the Asia-Pacific may be slowly coming to terms with the fact that states that promote people's security are more stable and prosperous, the region remains wary of unequivocally embracing the human security concept. States in the region are much more open to the "freedom from want" strand of human security (which is associated with Japan's advocacy, and linked to communal rights to development and poverty reduction) than the so-called Canadian "freedom from fear" strand (which is seen to embody more Western concerns for the protection of liberal, individual rights). To clarify this distinction, the region largely endorses a **nontraditional security (NTS)** discourse rather than a human security discourse. NTS refers to nonmilitary, transnational threats to peoples and states that require comprehensive (economic, social, and political) and multilateral remedies. NTS pertains to such problems as transnational organized crime, infectious disease, drug trafficking, people smuggling, environmental degradation, and natural disasters. Although NTS covers aspects of the human security agenda, the concepts differ in that NTS does not necessarily adopt the liberal, people-centered approach of human security concerned with individual safety and personal security.[10] Chapter 9 explores this difference in more detail.

The people-centered issues covered under the NTS rubric tend to be limited to food security, poverty reduction, natural disaster management, transnational crime, and environmental degradation. This is at the expense of "freedom from fear" issues, such as protecting civil and political rights and protecting civilians in armed conflict. Another tangible difference between human security and the Asia Pacific region's NTS concept is the absence of an express NTS commitment to human development, which understands development to be not merely a matter of economic growth but also about expanding people's empowerment and creating an environment in which people can enjoy long, healthy, and stable lives. The following section outlines how the failure to embrace a human security approach to economic development policy gives rise to new security challenges in the Asia-Pacific.

Development as a Human Security Challenge in the Asia-Pacific

During the past few decades, dynamic economic growth has created both new opportunities and new challenges for advancing human security in the Asia-Pacific region. Since the early 1980s, absolute poverty has declined faster than in any other region, household income has grown dramatically, and domestic economies have shifted from primarily industry and agriculture to manufacturing and services delivery. Strong and sustained economic growth has raised hundreds of millions of people out of crippling poverty. The population living below $1.25 per day fell from 53.9 percent in 1990 to 21.5 percent in about 2008, which amounted to 716 million fewer people living in extreme poverty.[11] Although poverty reduction is by and large a region-wide phenomenon, the record in some countries is truly staggering. Between 1990 and 2009, China lifted 27.7 million people each year from extreme poverty, amounting to more than half a billion people over the two-decade period (figure 12.1).[12]

While these statistics are undeniably impressive, many people in the Asia-Pacific continue to struggle to have their basic needs met. On account of its population size, the majority of the world's poor live in Asia. Extreme poverty levels remain a major challenge in some of the most populous states in the region, such as Bangladesh (43 percent of the population), India (33 percent), and Pakistan (21 percent).[13] Such trends, particularly in South Asia, contribute to the confronting fact that about 800 million Asians still survive on less than $1.25 per day and 1.7 billion live on less than $2 per day.[14] Progress in addressing child malnutrition and rural access to health services has been slow. Although significant headway has been made in reducing those who do not have access to safe water to 9 percent of the region's population, the same is not true for sanitation facilities; 45 percent of the region's people lack access to improved sanitation.[15]

Moreover, economic growth has not spread evenly. The Gini coefficient is a common measure of income inequality, with a higher Gini coefficient denoting greater inequality. Since the early 1990s, states that boasted some of the most impressive economic growth have also witnessed a significant bump in their Gini

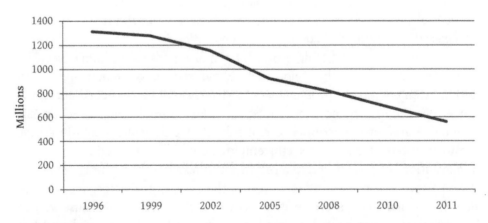

Figure 12.1 The Population of the Asia-Pacific Living on Less Than $1.25 per Day, 1996–2011

Note: Calculated by adding aggregations for East Asia and the Pacific and South Asia.

Source: PovcalNet, *Regional Aggregation Using 2005 PPP and $1.25/Day Poverty Line* (Washington, DC: World Bank, 2015), http://iresearch.worldbank.org/PovcalNet/index.htm?1.

coefficients; China jumped from 32 to 43, India from 33 to 37, and Indonesia from 29 to 39.[16] Such indicators suggest that behind aggregate economic growth and poverty reduction lay worsening conditions of inequality. According to the Asian Development Bank, the same catalysts of rapid economic growth—technological change, globalization, and market-oriented reform—are contributing to rising inequality in the Asia-Pacific.[17] This growth model privileges owners of capital, highly skilled workers, and urban development, and benefits the top echelon of income earners.

In the absence of people-centered policies and social safety nets to assist some of the most marginalized sectors and segments of society, women, minority groups, low-skilled workers, and the rural poor have not benefited from economic growth to the same degree as others.[18] In many cases, economic development has resulted in new human security threats among these populations. For example, the April 2013 collapse of an eight-story garment factory building near Dhaka, which killed 1,129 and injured approximately 2,515, brought into stark relief the insecurities facing workers' health and safety in the region. According to Human Rights Watch, the tragedy was "sadly predictable," due to egregiously poor textile industry standards that pay workers among the world's lowest wages, fail to ensure safe working conditions, and resort to threats and intimidation to thwart labor union activists from collectively bargaining for better working conditions.[19] Similar challenges to human security arise from governments colluding with large corporations to divest **subsistence farmers** of their land in order to develop large-scale plantations, coal mines, hydroelectric dams, and tourist resorts. Although this problem of land grabbing is prevalent throughout Southeast Asia, Cambodia has perhaps the worst record of development-forced displacement. Since 2003 widespread land grabbing has displaced an estimated 400,000 subsistence farmers in Cambodia, which has contributed to increased food insecurity among that country's rural poor.[20]

The Asia-Pacific model of development can also pose major human security challenges for women in the region, as rapid industrial development and urbanization often compound problems arising from deeply entrenched and acute gender discrimination and gender inequality. For example, men in South Asia have had a greater opportunity to exit subsistence agriculture and intensive livestock systems for service sector employment, which has led to a growing "feminization" of agriculture. As a result, South Asia has the highest proportion of women whose primary occupation is in the agriculture sector, and the employment of men in the services sector is twice that of women.[21] Because women in rural areas do not have the same legal access to land or financial credit as men, such trends have exacerbated food insecurity and retarded progress in addressing child malnutrition.[22] The point here is that economic development is gendered, and that aggregate growth in the Asia-Pacific has not automatically translated to gains in women's economic and political empowerment, or their well-being.[23] Without progress in upholding women's rights and overturning discriminatory norms and practices, the region's growth model may further diminish women's security.

Alongside the problems of rising inequality, poor labor conditions, rural displacement, and women's insecurity, the environment has also borne the cost of the region's path of economic growth. Asia leads the world in greenhouse gas emissions linked to global warming and climate change. Asian cities are among the most polluted in the world, and air pollution in the region, particularly in urban districts in China and India, is at deadly levels. In 2010, outdoor air pollution contributed to 1.2 million premature deaths in China, and 620,000 premature deaths in India (nearly 57 percent of the global total of premature deaths are due to poor air quality).[24] India's pollution rates recently surpassed China's, and India presently claims the unenviable position of the world's highest death rate because of chronic respiratory diseases.[25] Similarly, the "slash and burn" technique for clearing land for large-scale palm oil plantations has created a periodic "haze" crisis in Southeast Asia, in which residents of Brunei, Indonesia, Malaysia, Singapore, and southern Thailand experience hazardous levels of smog pollution.

Despite a widely held assumption in the region that economic growth makes states and peoples more secure, this analysis suggests that the growth model that has been adopted by many states in the Asia-Pacific may further exacerbate human insecurity. Without political accountability, transparency, gender equality, and people's empowerment, economic development can give rise to greater insecurity, endemic corruption, and severe abuse of power. Similar lessons are drawn in the following section, which examines the impact of armed conflict and mass atrocities in the region, and explores how advancing the Responsibility to Protect principle would help secure Asia-Pacific populations.

The Responsibility to Protect and Securing Asia-Pacific Peoples

State-sponsored violence and oppression in the Asia-Pacific have taken a tremendous toll on human life. In the late 1970s genocidal violence under the Khmer Rouge regime in Cambodia resulted in more than a third of the population being murdered and another third displaced. In Indonesia in 1965–66, anticommunist purges resulted in mass killings of an estimated 500,000 people. Indonesia's 1975–99 oppressive occupation of East Timor and the violence following East Timor's referendum for independence in 1999 resulted in an estimated 100,000 conflict-related deaths. More recently, the siege of Jaffna during the final brutal stage of the Sri Lankan civil war in 2009 resulted in upward of 300,000 victims of war crimes and crimes against humanity and a death toll of as many as 40,000 civilians.[26] North Korea's totalitarian regime is complicit in crimes against humanity to such a degree that the 2014 UN Commission of Inquiry on the human rights situation there attested that the "gravity, scale, and nature of these violations does not have any parallel in the contemporary world."[27] In Burma/Myanmar's Arakan state, the Buddhist monks and the local authorities, with the support of state security forces, have led a campaign of genocidal violence and ethnic cleansing against the Rohingya Muslim community.[28] To this day, state-led persecution and subnational conflicts throughout Burma/Myanmar and in southern Thailand, the Mindanao region of the Philippines, the West Papua region of Indonesia, the Chittagong Hill Tracts in Bangladesh, the Assam region of India, and elsewhere continue to imperil Asia-Pacific minority groups.[29] More alarming still, due in part to their highly militarized societies that have few constraints on executive power, Burma/Myanmar, Afghanistan, and Sri Lanka are among the top fifteen countries in the world forecasted to have the most acute risk of large-scale genocidal violence.[30]

Mitigating the scale and extent of violence against Asia-Pacific populations requires concerted state, regional, and international prevention and protection efforts. The Responsibility to Protect (R2P) principle is geared precisely to this end. Although R2P is sometimes mistakenly associated with a narrow focus on humanitarian intervention, the unanimous endorsement of R2P by UN member states at the 2005 World Summit defined R2P as a deeper and more comprehensive framework for preventing and responding to mass atrocities (i.e., widespread or systematic human rights violations perpetrated against large populations, often, but not always, in the context of armed conflict). According to the UN secretary-general, the General Assembly's 2005 endorsement of R2P entailed a policy commitment across three mutually reinforcing pillars: (1) each state's responsibility to protect its populations from **genocide**, **war crimes**, **ethnic cleansing**, and **crimes against humanity**; (2) the international community's responsibility to assist states in preventing these four crimes; and (3) the international community's responsibility to take timely and decisive action, through peaceful, diplomatic, and humanitarian means, and if necessary through other more forceful means, where a state manifestly fails to protect its population from these crimes.[31]

Since 2009 UN secretary-general Ban Ki-moon has issued annual reports on implementing the R2P principle, which have added conceptual clarity to R2P and translated UN member states' R2P commitment into the most comprehensive platform to date for preventing and responding to mass atrocities. The UN's R2P implementation agenda spans a range of preventive and remedial measures, including building state and regional capacity to prevent atrocities; "mainstreaming" atrocities prevention in UN institutions and agencies; building early warning and early response mechanisms; and fostering the political will and capacity to halt ongoing mass atrocities. Of particular importance, all these measures are part of the broader normative project of R2P: inculcating human protection as a defining attribute and goal of state sovereignty, and building acceptance and capacity within the UN Charter's framework for acting on a collective responsibility to protect populations from the massive and avoidable suffering resulting from widespread and systematic attacks against civilian populations.

Although the Asia-Pacific states remain wary of the potential for R2P to allow for greater military intervention and interference, most of these states have offered rhetorical support for the idea that a primary function of the state is to protect its population from mass atrocities and that regional and international institutions have a role in assisting in this enterprise.[32] The extent to which the region supports R2P therefore represents an important normative shift from a state-centered definition of sovereignty as an absolute right to noninterference, to a more people-oriented view of sovereignty as responsibility.

Given the region's strong commitment to noninterference and weak institutional architecture for protecting human rights, one of the key contributions that the R2P principle makes to securing Asia-Pacific populations is prodding regional leaders and institutions to respond to humanitarian crises within states. The regional response to Cyclone Nargis in Burma/Myanmar in 2008 offers a good example of how R2P amplified pressure for Burma/Myanmar and the regional authorities to take humanitarian action. Cyclone Nargis struck Burma/Myanmar on May 2, 2008, leaving in its wake more than 138,000 dead or missing and displacing some 800,000 people. The Burma/Myanmar military regime exacerbated the devastating humanitarian situation by obstructing international organizations and aid agencies from providing rapid relief and assistance to populations stranded without potable water, food, or shelter in the Irrawaddy Delta.

On the basis that the Burma/Myanmar authorities had demonstrated a lack of will or capacity to assist imperiled populations, French foreign minister Bernard Kouchner invoked R2P to press Burma/Myanmar's military leaders to allow unhindered humanitarian access for a comprehensive international emergency relief effort. By Jurgen Haacke's account, Western references to R2P served as a "rhetorical device" that compounded pressure on both the government of Burma/Myanmar and the Association of Southeast Asian Nations (ASEAN) to mount an effective relief effort. The invocation of R2P by leaders in the West pushed the Burma/Myanmar authorities to issue visas and grant access to international staff, and to receive an emergency assessment team from ASEAN. The heightened sense of urgency likewise catalyzed ASEAN into playing a mediating role between Naypyidaw and leaders in the West, and compelled the first-ever deployment of the ASEAN Emergency Rapid Assessment Team to provide assistance in humanitarian aid distribution.[33]

While the region's unprecedented effort to assist populations in the aftermath of Cyclone Nargis is noteworthy, it was carried out on an ad hoc basis. Indeed, despite international advocacy based on the R2P principle, the responses to other humanitarian crises in the region—such as those in Sri Lanka, North Korea, and Burma/Myanmar—have met with state and regional recalcitrance or inertia. Given deeply entrenched norms of noninterference and weak conflict management institutions, the most promising avenue for R2P to enhance human security in the region is through longer-term efforts aimed at (1) building normative support for the idea that massive human rights abuses within states are a legitimate regional and international concern and (2) building institutional capacity to take effective measures to prevent mass atrocities in the region.[34] In this vein, lessons can be drawn from the 2010–11 Council of Security Cooperation in the Asia-Pacific study group on R2P, which gathered regional experts to analyze the implications of R2P for regional actors and organizations.[35]

The study group comprised representatives from twenty countries, some of which (including China, Russia, and India) were represented by current or former diplomats or high-ranking military officials. The study group recommended that the ASEAN Regional Forum—the most inclusive security-oriented body in the region—play a key role in implementing R2P by establishing a Risk Reduction Centre to conduct early warning and assessment of the risk of mass atrocities; strengthening capacity to mediate conflicts before they escalate; and establishing a standing regional capacity to prevent and respond to atrocity crimes.[36] The R2P concept has also gained the support of high-profile advocates in the region, as witnessed in the 2014 report of the High-Level Advisory Panel on R2P in Southeast Asia (which was chaired by the former secretary-general of ASEAN, Surin Putsiwan, and was made up of a group of eminent senior Southeast Asian diplomats and policy advisers). The panel's report affirmed that the concepts of R2P "converge with ASEAN's vision of a peaceful, just, democratic, people-centered and caring community in Southeast Asia," and issued a number of recommendations for advancing atrocities prevention within existing ASEAN mechanisms and arrangements, such as the ASEAN Political Security Community, the ASEAN Intergovernmental Commission on Human Rights, the ASEAN Commission on the Promotion and the Protection of the Rights of Women and Children, and the ASEAN Institute for Peace and Reconciliation.[37]

Conclusion

This chapter has explored the merits of reorienting security in the Asia-Pacific from state security to human security to help address some of the most pressing threats to the survival and well-being of populations in the region. Human security places people at the heart of security matters, and emphasizes that state security relies on ensuring that people live free of extreme violence and deprivation.

In recent decades tremendous change has taken place in the Asia-Pacific, which has opened new avenues for promoting human security and has created new challenges for realizing its goals. Absolute poverty has declined in the region faster than in any other region, as the region's economic growth model has lifted millions of people out of crippling poverty. Alongside impressive economic growth, an amazing transformation took place in East and Southeast Asia in the

1980s as the region moved from one of the bloodiest, most war-prone regions to a region of relative interstate peace.[38] At first glance, this transformation upholds the widely held assumption that poverty and underdevelopment drive conflict and exacerbate insecurity, and therefore that economic growth makes states and peoples more secure. However, as this chapter has demonstrated, despite economic growth, populations in the region continue to face grave security threats brought on by poverty, inequality, and discrimination. Moreover, state security forces and paramilitary operatives continue to commit widespread and systematic violence against the region's civilian populations. The risk of large-scale genocidal violence—arguably, the most extreme violation of human security—remains a real concern.

Without a focus on human security, the benefits of economic growth remain elusive for many impoverished and marginalized segments of society throughout the region. Deeply entrenched and long-running subnational conflicts continue to gravely imperil indigenous and minority groups. The region's traditional security paradigm, which focuses on consolidating state power and rapid economic development, is unable to adequately address the widespread and acute problems associated with, inter alia, rising inequality, dangerous working conditions, extreme poverty, gender-based discrimination and gender inequality, state-sponsored violence, deep-seated communal conflict, and environmental degradation. On both normative and practical levels, such festering problems lay bare the need to adopt a more comprehensive, people-centered approach to security that safeguards the region's most vulnerable populations.

Key Points

- Human security places individuals as the primary referent object of security, diminishes the significance of international borders, and aims to advance the survival and dignity of human beings.
- Definitions of human security differ between those who embrace a broad definition that includes human development and meeting basic human needs (i.e., freedom from want), versus others who adopt a narrower definition concerned with protecting populations from conflict-related violence (i.e., freedom from fear).
- Economic growth has raised millions of people out of extreme poverty, but it has also created new human security challenges. A human development approach is needed to comprehensively address these security challenges.
- The international community endorsed the R2P principle in 2005. It has amplified pressure on states to take action on humanitarian crises in the region and has been an impetus for shifting understandings of sovereignty in the region and building institutional capacity to prevent and respond to risks of atrocities in the future.

Questions

1. How does human security differ from traditional concepts of security?

2. Why has the Asia-Pacific region been slow to embrace human security?

3. Why is development a security challenge in the Asia-Pacific?

4. How might the R2P principle help secure Asia-Pacific populations?

5. What is the benefit of reorienting the understanding of security in the Asia-Pacific to focus on people rather than, or along with, states?

Guide to Further Reading

Acharya, Amitav. "Human Security: East versus West." *International Journal* (Toronto) 56, no. 3 (2001): 442–60. http://dx.doi.org/10.2307/40203577.
This article outlines some of the challenges to advancing human security in Asia, with a focus on the norms and principles that undergird the Asia-Pacific security context.

Annan, Kofi. *In Larger Freedom: Toward Development, Security, and Human Rights for All.* United Nations General Assembly Document A59/2005. New York: United Nations, 2005.
This secretary-general's report is a good example of UN efforts to link security, development, and the promotion of human rights, and is worth reading as an introductory text on the human security policy discourse.

Bellamy, Alex J. *Responsibility to Protect.* Cambridge: Polity, 2009.
This book offers the most comprehensive account of the evolution of the R2P concept leading to its adoption at the 2005 World Summit, and outlines practical steps for achieving the aims of R2P to protect populations from atrocity crimes.

Caballero-Anthony, Mely. "Revisioning Human Security in Southeast Asia." *Asian Perspective* 28 (2004): 155–89.
This article, written by perhaps the leading academic on human security in the region, explores how civil society organizations have advanced human security concepts in Southeast Asia, despite deeply entrenched norms of noninterference.

Kraisoraphong, Keokam, ed. "Thailand and the Responsibility to Protect." *Pacific Review* (Special Issue) 25, no. 1 (2012). http://dx.doi.org/10.1080/09512748.2011.632960.
This special issue includes articles from leading academics and practitioners in the region on some of the challenges and opportunities for advancing the R2P principle and human protection through Asia-Pacific regional institutions.

Notes

1. United Nations Development Program, *Human Development Report* (New York: United Nations, 1994), 23.

2. Amitav Acharya, "Human Security: East versus West," *International Journal* 56 (2001): 446–48.

3. Mely Caballero-Anthony, "Revisioning Human Security in Southeast Asia," *Asian Perspective* 28 (2004): 160–61.

4. Rizal Sukma, "The ASEAN Political and Security Community (APSC): Opportunities and Constraints for the R2P in Southeast Asia," *Pacific Review* 25 (2012): 140.

5. David Capie and Paul Evans, *The Asia Pacific Security Lexicon*, 2nd ed. (Singapore: Institute of Southeast Asian Studies, 2002).

6. Amitav Acharya, *Promoting Human Security: Ethical, Normative, and Education Frameworks in South East Asia* (Paris: UNESCO, 2007), 12.

7. Dewi Fortuna Anwar, "Human Security: An Intractable Problem in Asia," in *Asian Security Order: Instrumental and Normative Features*, ed. Muthiah Alagappa (Stanford, CA: Stanford University Press, 2003), 537.

8. Acharya, *Promoting Human Security*, 23.

9. Acharya, "Human Security," 565.

10. Melissa Curley, "Human Security's Future in Regional Cooperation and Governance?" *Australian Journal of International Affairs* 66 (2012): 529.

11. Asian Development Bank, *Asian Development Outlook 2012: Confronting Rising Inequality in Asia* (Manila: Development Asia, 2012), 38.

12. Asian Development Bank, *Key Indicators for Asia and the Pacific* (Manila: Development Asia, 2013), 99.

13. Ibid.

14. Ibid.

15. Ibid., 150.

16. Asian Development Bank, *Deepening Divide: Can Asia Beat the Menace of Rising Inequality?* (Manila: Development Asia, 2013), 10.

17. Ibid.

18. Asian Development Bank, *Asian Development Outlook 2012*, 37.

19. Human Rights Watch, "Bangladesh: Tragedy Shows Urgency of Worker Protections," April 25, 2013, http://www.hrw.org/news/2013/04/25/bangladesh-tragedy-shows-urgency-worker-protections.

20. John Cherry, "The Great Southeast Asian Land Grab," *The Diplomat*, http://thediplomat.com/2013/08/the-great-southeast-asian-land-grab/.

21. Asian Development Bank, *Gender Equality and Food Security: Women's Empowerment as a Tool against Hunger* (Manila: Development Asia, 2013), 20–21.

22. Ibid.

23. Jacqui True, Sara Niner, Swati Parashar, and Nicole George, *Women's Political Participation in Asia and the Pacific* (New York: Conflict Prevention and Peace Forum of United Nations Social Science Research Council, 2012), 12.

24. Edward Wong, "Air Pollution Linked to 1.2 million Premature Deaths in China," *New York Times*, April 1, 2013, http://www.nytimes.com/2013/04/02/world/asia/air-pollution-linked-to-1-2-million-deaths-in-china.html.

25. Gardiner Harris, "Beijing's Bad Air Would Be Step Up for Smoggy Delhi," *New York Times*, January 25, 2014, http://www.nytimes.com/2014/01/26/world/asia/beijings-air-would-be-step-up-for-smoggy-delhi.html?_r=0.

26. United Nations, *Report of the Secretary-General's Panel of Experts on Accountability in Sri Lanka* (New York: United Nations, 2011).

27. UN Human Rights Commission, *Report of the Commission of Inquiry on Human Rights*

in the Democratic People's Republic of Korea, General Assembly Document A/HRC/25/63/7 (New York: United Nations, 2014), 15.

28. Human Rights Watch, "All You Can Do Is Pray: Crimes against Humanity and Ethnic Cleansing of Rohingya Muslims in Burma's Arakan State," April 22, 2013, http://www.hrw .org/reports/2013/04/22/all-you-can-do-pray-0.

29. MAR Project 2014, http://www.cidcm.umd.edu/mar/.

30. Charles R. Butcher and Benjamin E. Goldsmith, Dimitri Semenovich, and Arcot Sowmya, "Political Instability and Genocide in the Asia-Pacific: Risks and Forecasts," August 2012, http://sydney.edu.au/arts/research/atrocity_forecasting/downloads/docs /GenocideForecastingReportAsia_2012afp.pdf.

31. United Nations, "World Summit Outcome Document," General Assembly Document A/60/L.1, September 2005, paras. 138–40; Ban Ki-moon, *Implementing the Responsibility to Protect: Report of the Secretary-General*, General Assembly Document A/63/77 (New York: United Nations, 2009).

32. Alex J. Bellamy and Sara E. Davies, "The Responsibility to Protect in the Asia-Pacific Region," *Security Dialogue* 40 (2009): 547–74.

33. Jurgen Haacke, "Myanmar, the Responsibility to Protect, and the Need for Practical Assistance," *Global Responsibility to Protect* 1 (2009): 156–84, at 170.

34. Sukma, "ASEAN Political and Security Community"; Herman Kraft, "RtoP by Increments: The AICHR and Localizing the Responsibility to Protect in Southeast Asia," *Pacific Review* 25 (2012): 27–49; Anthony Mely-Caballero, "The Responsibility to Protect in Southeast Asia: Opening Up Spaces to Advance Human Security," *Pacific Review* 25 (2012): 113–34.

35. Council of Security Cooperation in the Asia Pacific, "Study Group on the Responsibility to Protect: Final Report," 2011, http://www.cscap.org/index.php?page =responsibility-to-protect.

36. Ibid., 22.

37. High-Level Advisory Panel, "Mainstreaming the Responsibility to Protect in Southeast Asia: Pathway towards a Caring ASEAN Community," 2014, available at http:// r2psoutheasasia.org/.

38. Alex J. Bellamy, "The Other Asian Miracle? The Decline of Mass Atrocities in East Asia," *Global Change, Peace, and Security* 26, no. 1 (2014): 1–19.

Conclusion: What Is the Asia-Pacific's Likely Security Future?

Brendan Taylor and William T. Tow

Reader's Guide

This chapter concludes the book by examining four major models that have been applied to assess the politics of the security order in the Asia-Pacific. The first three fall under the neorealist approach. The first model, "hegemony," refers to an order where one country is dominant and where the other countries largely accept its dominance. The second, "strategic condominium," is present when great powers agree to share power with one another. The third, a "balance-of-power system," describes an order that is expected to emerge from the equilibrium or "balance" generated by competition between the key players. The fourth and final model, "institutionalism," comes from the neoliberal approach and describes a situation in which multilateral organizations become the main instruments for addressing challenges and often for managing the actions of the larger powers. The key message to take away from this chapter is that, although none of these models fully captures the current dynamics of Asia-Pacific security, the region's most likely security future can best be anticipated by drawing upon elements of each of them.

Introduction

Scholars, policymakers, and commentators on security politics are notorious for describing the era in which they are writing as being the most "complex" in history. As the contributions to the current volume illustrate, however, those writing in the current era have an actual basis for making such a claim. To be sure, there have arguably been other, much more *dangerous*, eras. During the Cold War, for instance, the world constantly hovered at the edge of the nuclear precipice. That was a threshold that, if crossed, would have meant the end of civilization as we know it. Yet in terms of the sheer range of strategic risks, challenges, and opportunities that the Asia-Pacific currently faces, it is difficult to recall a period in history when such a varied number of threats and challenges have coexisted, and in some instances converged, to the extent we see today. As the world's economic and military center of gravity shifts decisively to the region, the complexities of international relations are nowhere more apparent than in this increasingly critical region.

This chapter concludes the book's analysis by trying to make some value judgments about how the different actors (chapters 1–5), issues (chapters 6–10), and possible solutions (chapters 11 and 12) will shape the security environment of the Asia-Pacific. As the analysis contained in this volume makes clear, scholars and practitioners continue to grapple with the question of what kind of security order will ultimately emerge from this menagerie of competition and cooperation. Writing a decade ago, the doyen of Asia-Pacific security studies, Muthiah Alagappa, observed that three predominant conceptions of security order had been applied to this part of the world—hegemony, strategic condominium / balance of power, and institutionalism.[1] In this final chapter, we consider the Asia-Pacific's likely security future through each of these conceptual lenses. We observe that, though each of these models is able to capture elements of the region's current and evolving security dynamics, none is able to account for these in their entirety. We therefore conclude that understanding the region's security is impeded by incomplete paradigms and that more sophisticated typologies are required to accommodate the complex range of threats, challenges, and possible solutions that are drawn out by this book's contributors.

Hegemony

Hegemony, the first conception of security order to which Alagappa refers, envisages that order is preserved by a dominant state in the system. This "hegemon" exercises primacy due to both its possession of superior military and economic resources and it willingness to use these attributes. A further condition of a hegemonic system is that the other states in this system ultimately need to accept the leadership of the dominant state. Typically, they will accede to its leadership—often by forming alliances, or "bandwagoning," with the dominant power—due to the hegemon's willingness and ability to provide military and economic "public goods," which in turn assist with promoting systemic stability.[2]

Some commentators have argued that the United States has exercised a form of "benign hegemony" in the Asia-Pacific. As reflected by Brad Glosserman's comments in chapter 1, this school of thought has viewed the United States as serving as a stabilizing influence in this region for much of the period since the end of World War II in 1945. The United States has used a number of instruments of statecraft to exercise leadership during this time. It has maintained major military bases—for instance, in Japan and South Korea—as well as sheltering its friends and partners beneath its so-called nuclear umbrella. These actions have arguably contributed to regional stability by diminishing the need for US allies such as Japan, South Korea, and Australia to develop their own nuclear arsenals. The United States has also maintained its leadership through the use of economic incentives, such as the preferential trading arrangements it made available to some of its newfound allies at the birth of the San Francisco System and again in the early to middle 2000s as a reward for their support during the so-called War on Terror.[3] The United States has further endeavored to exercise and maintain leadership through the use of "soft power"—meaning a country's power of attraction derived primarily from its culture, values, and foreign policy.[4] However, it has also shown a willingness to use force to preserve its leadership and the stability of the region. It led the Western effort in the Korean War and Vietnam War. The United States was also the "first responder" to major traditional and nontraditional regional crises, such as conflict on the Korean Peninsula during the years 1950–53, the Taiwan Strait crisis of 1995–96, the Boxing Day tsunami of December 2004, or Japan's "triple disaster" of March 2011. The US pivot can be viewed as an effort to strengthen Washington's hegemonic role in the Asia-Pacific.

A growing number of commentators are, however, questioning the United States' capacity to maintain this role—one it has arguably enjoyed since the end of World War II—in the future. They point to an American economy increasingly under strain, particularly in the aftermath of the global financial crisis of 2008–9. They also cite an American public whose appetite for military deployments in far distant theaters has been severely dampened by difficult and protracted operations in Iraq and Afghanistan. Perhaps most important, those insisting that America is in decline point out that power relativities are shifting in the Asia-Pacific region as Asia's giants—the People's Republic of China (hereafter, China) and India—stand up and as so-called middle powers such as Indonesia, South Korea, and Vietnam begin to emerge as more important strategic actors in their own right. This is not to deny the reality that the United States still spends significantly more on its military than all these actors combined. Rather, this school of thinking is a reflection of the historical inevitability that military weight eventually follows economic weight.[5] According to some estimates, China will overtake the United States as the world's leading economy (in purchasing power parity).[6]

Do these trends suggest that China will, in time, also supplant the United States as an Asia-Pacific hegemon? Some scholars, such as John Mearsheimer, have argued that regional hegemony has traditionally been a natural aspiration for every rising great power and that China is no exception in this regard.[7] On the back of its growing economic weight, China has been increasing its defense expenditures and has been buying a range of more powerful weapons platforms for more than two decades now. Since 2009 Beijing's so-called new **diplomacy**—a kinder, gentler, more nuanced foreign policy approach—of the late 1990s and most of the 2000s appears to have gone by the wayside. In its place, Beijing has become much more assertive in most if not all of its key foreign relationships—particularly in relation to the Asia-Pacific's territorial disputes in the East China Sea and South China Sea, as discussed both by Lowell Dittmer in chapter 2 and by James Manicom in chapter 7. Indeed, some commentators have expressed concern that China's "nine-dash-line" outlining its claims in the South China Sea is representative of a new Chinese sphere of influence and the outline of a new era of Chinese hegemony over Southeast Asia and potentially beyond.[8] In delivering his landmark speech to the Conference on Interaction and Confidence-Building Measures in Asia meeting in Shanghai in May 2014, Chinese president Xi Jinping insisted that "it is for the people of Asia to run the affairs of Asia, solve the problems of Asia, and uphold the security of Asia," and that the entrenchment of military alliances is not conducive to the region's security order.[9]

Yet just as some commentators question the viability of continued American hegemony in the Asia-Pacific, so do others doubt the capacity of China to truly become the dominant state in this part of the world. As discussed in the next section of this chapter, that is partly due to the unwillingness of many other regional states to accept Chinese leadership. But more important, China faces a raft of internal challenges and constraints—including corruption, dire environmental problems, a growing gap between rich and poor that is causing deep resentment and unrest, homegrown terrorism, and unfavorable demographic trends caused by its one-child policy, to name just a few. In combination, these challenges could well slow China's rise considerably, if not ultimately prevent it from becoming a regional hegemon in any true sense of the term. Moreover, as Dittmer argues in chapter 2, China is also itself aware of the risks of exercising hegemony. Certainly in response to regional crises over recent years—from the Boxing Day tsunami of 2004 to Typhoon Haiyan in 2013—Chinese willingness and/or ability to lead has been notable by its absence. As one analyst concluded at the time of the latter disaster, on the question of China's regional leadership, "whether it is about soft power or the effective deployment of military force, Beijing still has a long way to go."[10]

Strategic Condominium / Balance of Power

The second conception of security order to which Alagappa refers, a strategic condominium, has been a relatively rare occurrence in the history of international relations. As Carsten Holbraad once observed, "Looking to history for instruction about the nature of such condominia, one does not find a host of examples. Cases of dualistic situations are numerous, but generally characterized by rivalry between the two dominant powers."[11] Although Holbraad was only able to find three such instances in his study of modern history, speculation intensified in the wake of the 2008–9 global financial crisis that a new US-China condominium was emerging. Respected figures such as Henry Kissinger and Zbigniew Brzezinski suggested that a new "Group of 2" could be established if Sino-US economic cooperation was translated into the strategic realm to tackle a raft of international problems, including climate change, India–Pakistan tensions, the Arab-Israeli dispute, nuclear proliferation, and conflict in **failed states**.[12] At various times, it has also been fashionable to advocate a condominium (or concert of powers), where more than two players share power on a largely informal basis, along much the same lines as the nineteenth-century Concert of Europe.[13]

Such analyses remain in the minority, however, and there is still a good deal of skepticism that the United States and China will ultimately be able to "share power" in any meaningful sense. This is partly due to the significant normative and geopolitical differences that China and the United States exhibit, as illustrated by Xi Jinping's statement cited above and by the Obama administration's initiation of its "rebalancing" or "pivot" strategy during its first term in office, which has been largely viewed as countering Chinese policies and influence in the region.[14] Writing critically of the Group of 2 concept when it first emerged in the late 2000s, the American analysts Elizabeth Economy and Adam Segal observed that the "lack of US-China cooperation does not stem from a failure on Washington's part to recognize how much China matters, nor is it the result of leaders ignoring the bilateral relationship. It derives from mismatched interests, values, and capabilities."[15] At the same time, however, it is the similarities between China and the United States as much as their differences that make the power-sharing ideal so difficult for them to implement in practice. Both countries have long seen themselves and their societies as somehow "exceptional" and unique, superior even to the other members of the international system—one another included. Since its founding, for instance, the United States has viewed itself as a "shining city on the hill" for other societies to emulate.[16] Equally, the Chinese call their country Zhong Guo (meaning "middle country"), a still relatively modern idea that harks back in some respects to the central position that ancient China occupied during the era of the so-called tributary system.[17] Such deep-seated exceptionalist mindsets mean that the notion of power sharing does not come readily to either Beijing or Washington.[18] Here we see the value of constructivist theories, which ask us to look beyond material issues to the role of ideas in shaping security behavior.

If an essentially "managed balance" in the form of a "strategic condominium" has its shortcomings, what of the prospects for a more unregulated order based on a balance of power? The balance-of-power conception to which Alagappa refers posits that the cooperative management of great power relations as envisaged under a power condominium is not a feasible approach to security order, given the inherently anarchic nature of international relations. Instead, as China's material power gradually continues to rival the dominance of the United States, this conception—favored by those of a realist outlook—predicts that these two powers will and, indeed, should seek to balance against one another.[19] They will seek to do so in at least one of two ways. First, they will aim to increase their own military capabilities (an approach referred to as "internal balancing"). And second, they will endeavor to enhance their capabilities by developing further strategic alignments (what is known as "external balancing"). More specifically, the United States will seek to reinforce its traditional postwar bilateral alliance system in the Asia-Pacific while China will attempt to weaken it through various diplomatic and strategic means. In theory at least, provided that these competitive processes result in the two dominant powers in the system largely offsetting one another and arriving at a roughly approximate balance of power, order and stability can then be maintained. Again, European examples—in this case Europe in the sixteenth, seventeenth, eighteenth, and twentieth centuries—are often cited as indicative of what might transpire in the Asian security order during the twenty-first century.[20]

While essentially European in origin, the application of the balance-of-power concept to an Asian setting is not new. Writing more than a hundred years ago, for instance, the historian and strategist Alfred Thayer Mahan argued that "US security was tied to the balance of power in Asia as well as Europe."[21] A flurry of writings referring to an "Asian balance of power" appeared in the 1960s and 1970s. A trio of papers published in 1967–68 under the auspices of the London-based International Institute for Strategic Studies applied the concept and analyzed it from distinct American, European, and Australian perspectives.[22] In the early 1970s Hedley Bull published an oft-cited essay in the prominent American policy journal *Foreign Affairs* in which he analyzed the emergence of what he termed a "complex" new balance of power in Asia and the Pacific.[23] In another widely cited work, published during the mid-1990s, again under the auspices of the International Institute for Strategic Studies, the Australian academic and former senior government official Paul Dibb also addressed the emergence of a "new" Asian balance of power.[24]

Despite this long and distinguished lineage, it has become increasingly fashionable since the end of the Cold War to disparage the application of the balance-of-power concept to the Asia-Pacific. A common refrain among scholars has been that this European construct is inapplicable to the highly variegated Asian region.[25] Consistent with this, scholars have argued that a noticeable lack of balancing behavior has been evident in the Asia-Pacific for the better part of a quarter century. For instance, a leading scholar of Asian security, David Kang, asks in one of his best-known books why East Asian countries have accommodated rather than balanced the rise of China.[26] The Harvard University professor Alastair Iain Johnston agrees, observing in his review of the scholarship on Asian security that much of this body of work "rejects this structural take on the region." In his terms, "most East Asian states are not seen as balancing against China. Nor are they balancing against the United States, as they should be if material power differentials matter."[27] Another recent, prominent example of such work is Steve Chan's latest book *Looking for Balance*, in which he attributes the nonoccurrence of balancing behavior in the Asian context to the fact that "East Asian elites have collectively pivoted to a strategy of elite legitimacy and regime survival based on economic performance rather than nationalism, military expansion, or ideological propagation."[28] In other words, according to Chan, butter has taken precedence over guns, plowshares over swords.

While this logic has continued to hold up fairly well for much of the past two and a half decades, there are indications that it is beginning to break down and become less convincing as signs of strategic competition emerge in the Asia-Pacific region. As previously noted, America's Asian "pivot" or "rebalancing" strategy has been viewed by many as driven by a perceived need to check China's growing power and influence in this part of the world. Japan, too, has tightened its alliance with the United States and is currently in the midst of shedding long-standing constitutional constraints on the use of its military, largely in response to China's rise. India appears increasingly comfortable with balance-of-power politics—including in the form of deepening security ties with Japan—where previously, as Envall and Hall detail in chapter 3, it had avoided such approaches in favor of a nonaligned strategic posture. Even in Southeast Asia, where many if not most countries have preferred "hedging" strategies involving the maintenance of equidistance between the region's larger powers, middle power and smaller countries such as Vietnam and the Philippines are also showing the beginnings of adopting more hard-line postures.[29]

To be sure, any balancing behavior that is becoming apparent in the Asian region remains highly varied across countries and thus far has tended to be of the "softer," more indirect variety. However, if strategic competition between the region's major players—particularly the United States and China, but also Japan and increasingly India—continues to intensify, the possibility of harder, more traditional forms of balancing behavior beginning to emerge cannot be discounted.[30]

Institutionalism

The third and final conception of Asian security order outlined by Alagappa draws from a more liberal perspective and is known as institutionalism. As its name implies, multilateral *institutions* are central to this conception of security order, both as mechanisms for building trust and transparency and as a potential means for diluting the power of the more dominant powers in the system. For this reason, the so-called middle and small powers analyzed by Andrew Carr (in chapter 4) and Joanne Wallis (in chapter 5) have tended to be the champions of this conception of security order. In the Asian setting, the Association of Southeast Asian Nations (ASEAN) is cited by some commentators as illustrative of what can be achieved by institutionalism at the subregional (i.e., Southeast Asian) level.[31] ASEAN has also positioned itself to occupy the "driver's seat" in Asia's emerging security architecture by playing a central position in many of the region's leading multilateral institutions—including the East Asia Summit, the ASEAN Regional Forum, and the ASEAN Defense Ministers' Meeting Plus process. Through these multilateral mechanisms, the smaller and medium-sized powers of Southeast Asia aspire to create a security order by "enmeshing" the larger powers and shaping their behavior by exposing them to ASEAN principles, rules, and norms.[32]

Asian institutionalism is a surprisingly new phenomenon. At the end of the Cold War, there were very few channels for multilateral dialogue in the Asia-Pacific, and the prospects for cooperation of this nature appeared bleak. Beijing, for example, was highly suspicious that such mechanisms could be used by smaller and middle powers to essentially "gang up" against China and to constrain its growing regional influence—precisely as institutionalist theories suggest. Likewise, Washington was equally apprehensive at the prospect that new multilateral mechanisms could be used to undermine the centrality of the United States-led network of Asian alliances to the region's security and, over time, to perhaps even eliminate the need for this structure altogether. It is somewhat ironic, therefore, that a deepened commitment to Asian multilateralism has been one of the main policy innovations of the US pivot toward Asia.[33]

As Mathew Davies discusses in chapter 11, channels for multilateral dialogue have certainly proliferated since the early 1990s, to the point where there are now literally hundreds of such mechanisms at the so-called track one (or official) and track two (nonofficial) levels. That said, the performance of these institutions has been marginal at best. Particularly disappointing has been the lack of capacity by this multitude of instrumentalities to respond to crises in the region when these have emerged. One of the striking characteristics across a raft of crises—including the North Korean nuclear crisis of the early 1990s; the Taiwan Strait crisis of 1995–96; the Asian financial crisis of the late 1990s; major natural disasters, such as the 2004 Indian Ocean tsunami and Typhoon Haiyan of 2013; and the intensification of territorial disputes in the East China Sea and South China Sea, which continue today—has been the inability of the region's multilateral institutions to provide any kind of collective and effective response to these contingencies. This makes it unlikely that institutionalism can solve some of the intractable regional security crises, such as maritime disputes (chapter 7), terrorism (chapter 8), cyber issues (chapter 10), or even changing attitudes to encourage a human security focus (chapter 12).

Indeed, rather than facilitating enhanced cooperation of the kind that could be used to address such regional crises as and when they occur, one worrying trend in recent years has been the tendency for Asia's great powers to play out their emerging competition through a number of these groupings. China's strong backing for the Shanghai Cooperation Organization, the ASEAN+3 process, and the previously mentioned Conference on Interaction and Confidence Building Measures in Asia, for instance, stems from the fact that none of these organizations includes the United States as a formal member.[34] Each therefore serves one of China's purported longer-term strategic aims of marginalizing American influence in the Asia-Pacific. Likewise, Tokyo's backing of more inclusive groupings such as the East Asia Summit and its deliberate efforts to expand the membership of such groups can potentially be seen in the context of their deepening strategic competition with China, with such organizations serving as vehicles to dilute Chinese power and influence rather than further the cause of regional cooperation.

A Mosaic of Models?

This concluding chapter has demonstrated that elements of each of the models of security order that Alagappa identified a decade ago are apparent in contemporary Asia-Pacific security politics. Yet none among these models appears dominant; nor are they likely to become so in the foreseeable future. Neither the United States nor China appears likely to unequivocally occupy the position of the region's hegemon, and they are unlikely to be able to share power, as the notion of a strategic condominium suggests. While elements of balance-of-power politics can be said to be reemerging after a relatively lengthy hiatus, this still remains a work in progress. And as the analysis of Asian institutionalism provided in this chapter suggests, one needs to be wary of straight-line extrapolations and of assuming that modes of strategic behavior must inevitably move beyond the nascent form, as Asia's multilateral organizations—though plentiful—have thus far arguably failed to do.

With none of the traditional pathways to security order that Alagappa identified able to claim dominance, the future of Asia-Pacific security since the US pivot strategy can best be explained and understood by combining aspects of each of these traditional models. A leading analyst of Asian security, David Shambaugh, hinted at this approach some years back when he suggested employing "a mosaic of models," given that "one size does not, and cannot, fit all in a region as diffuse and diverse as Asia."[35] More recently, another leading scholar of Asian security, Amitav Acharya, has embraced a similar approach, proposing what he terms a "consociational security order" that draws from various theoretical lenses, including realism, institutionalism, and consociational theory from the field of comparative politics.[36] In the final analysis, the very diversity of the strategic dynamics, security challenges, and potential solutions addressed in this volume suggests that understanding how such approaches intersect will be increasingly central to the shaping and management of the Asia-Pacific's security future.

Key Points

- The Asia-Pacific faces a vast array of security challenges. They involve disputes between actors (both state and nonstate) and disputes over issues that threaten some or all of the states and people in the region.
- In a heterogeneous environment like the Asia-Pacific, it is unlikely that any one solution will be able to overcome all the problems the region faces. This

suggests something of the limitation of security study theories, given their claims to universal assumptions about international security.

- A mosaic-of-models approach, which begins with this recognition of limitations, might be the best that can be hoped for.
- Clearly identifying what solutions might be applicable and making predictive judgments about what will happen is a risky process, but it carries analytical value by forcing students of security to be clear about how they weigh the evidence of current events and how they determine what is most important.

Questions

1. What are the most important security challenges facing the Asia-Pacific?

2. Can any of the actors listed in this book be ignored when seeking solutions to the security issues of the Asia-Pacific?

3. Which security study theory best explains the challenges faced by the Asia-Pacific and the solution?

4. How do the various subregions of the Asia-Pacific differ in their likely security futures?

5. Should one be pessimistic or optimistic about the security future of the Asia-Pacific?

Further Reading

Alagappa, Muthiah, ed. *Asian Security Order: Instrumental and Normative Features*. Stanford, CA: Stanford University Press, 2003.
A classic work in analyzing the emerging security architecture of the Asia-Pacific. This volume helped established Alagappa as one of the doyens of Asia-Pacific security studies.

Dibb, Paul. *Towards a New Balance of Power in Asia*. Adelphi Paper 295 (Oxford: Oxford University Press for International Institute for Strategic Studies, 1995).
A major assessment of the changing balance of power in Asia that still rewards careful readers today. Makes some important judgments about the evolving region's security that have proven impressively prescient.

Kennedy, Paul. *The Rise and Fall of the Great Powers*. New York: Random House, 1987.
The classic work on the role of the great powers and the significance of economic change for influencing strategic reorientation and the prospects of war. Still extremely relevant, despite being published nearly thirty years ago.

Tow, William T. *Asia-Pacific Strategic Relations: Seeking Convergent Security*. Cambridge: Cambridge University Press, 2001.
Argues for a synthesis approach that merges liberal and realist approaches, advocating that a "convergent security" framework be adopted to create enduring regional security.

Notes

1. Muthiah Alagappa, "Constructing Security Order in Asia: Conceptions and Issues," in *Asian Security Order: Instrumental and Normative Features*, ed. Muthiah Alagappa (Stanford, CA: Stanford University Press, 2003), 72.

2. Steven Walt, *The Origin of Alliances* (Ithaca, NY: Cornell University Press, 1987), 17–21, 27–32.

3. Kent E. Calder, "Securing Security through Prosperity: The San Francisco System in Comparative Perspective," *Pacific Review* 17, no. 1 (March 2004): 135–57.

4. For further reading, see Joseph S. Nye, "What China and Russia Don't Get about Soft Power," *Foreign Policy* 29 (April 2013), http://www.foreignpolicy.com/articles/2013/04/29/what_china_and_russia_don_t_get_about_soft_power.

5. The classic text here is by Paul Kennedy, *The Rise and Fall of the Great Powers: Economic Change and Military Conflict from 1500 to 2000* (New York: Random House, 1987).

6. See, e.g., Chris Giles, "China Poised to Pass US as World's Leading Economic Power This Year," *Financial Times*, April 30, 2014.

7. John J. Mearsheimer, *The Tragedy of Great Power Politics* (New York: W. W. Norton, 2001).

8. See, e.g., Denny Roy, *Return of the Dragon: Rising China and Regional Security* (New York: Columbia University Press, 2013).

9. "Statement by H. E. Mr. Xi Jinping, President of the People's Republic of China," http://www.fmprc.gov.cn/mfa_eng/zxxx_662805/t1159951.shtml.

10. Rory Medcalf, "China Throws Away a Chance to Lead," *Wall Street Journal Asia*, November 15, 2013.

11. Carsten Holbraad, "Condominium and Concert," in *Super Powers and World Order*, ed. Carsten Holbraad (Canberra: Australian National University Press, 1971), 4.

12. See Henry Kissinger, "The Future of US-China Relations: Conflict Is a Choice, Not a Necessity," *Foreign Affairs* 91, no. 2 (March–April 2012), 44–55; and Zbigniew Brzezinski, "The Group of Two That Could Change the World," *Financial Times*, January 14, 2009.

13. The classic text on the concert of Europe is by Henry Kissinger, *A World Restored: Metternich, Castlereagh, and the Problems of Peace 1812-1822* (London: Weidenfeld & Nicolson, 1957). Recent discussions on prospects for an Asian concert include those by Amitav Acharya, "A Concert of Asia?" *Survival* 41, no. 3 (Autumn 1999): 84–101; Douglas T. Stuart, "Towards Concert in Asia," *Asian Survey* 37, no. 3 (March 1997): 229–44; and Sandy Gordon, "The Quest for a Concert of Powers in Asia," *Security Challenges* 8, no. 4 (Autumn 2012): 35–55.

14. See Kenneth Lieberthal, "The American Pivot toward Asia: Why President Obama's Turn to the East Is Easier Said Than Done," *Foreign Policy*, December 21, 2011, http://www.foreignpolicy.com/articles/2011/12/21/the_american_pivot_to_asia. Lieberthal noted that the pivot strategy is designed to do much more than merely confront the Chinese but observed that the US media portrayed it as just that and noted that the new posture "in many ways reinforced China's abiding suspicions about the United States."

15. Elizabeth C. Economy and Adam Segal, "The G-2 Mirage: Why the United States and China Are Not Ready to Upgrade Ties," *Foreign Affairs* 8, no. 3 (May–June 2009): 15.

16. Walter Russell Mead, *Special Providence: American Foreign Policy and How It Changed the World* (New York: Routledge, 2002).

17. David C. Kang, *East Asia before the West: Five Centuries of Tribute and Trade* (New York: Columbia University Press, 2010).

18. Nevertheless, this is a policy course advocated by Hugh White, *The China Choice: Why America Should Share Power* (Collingwood, Australia: Black Inc., 2012).

19. See, e.g., Aaron Friedberg, *A Contest for Supremacy: China, America and the Struggle for Mastery in Asia* (New York: W. W. Norton, 2011).

20. For further reading, see Hugh White, "Why War in Asia Remains Thinkable," *Survival* 50, no. 6 (December 2008–January 2009): 85–104.

21. Alfred Thayer Mahan, *The Problem of Asia: Its Effect upon International Politics* (New Brunswick, NJ: Transaction, 2003).

22. William Chapin, *The Asian Balance of Power: An American View*, Adelphi Paper 35 (Oxford: Oxford University Press for International Institute for Strategic Studies, 1967); Michio Royama, *The Asian Balance of Power: A Japanese View*, Adelphi Paper 42 (Oxford: Oxford University Press for International Institute for Strategic Studies, 1967); Coral Bell, *The Asian Balance of Power: A Comparison with European Precedents*, Adelphi Paper 44 (Oxford: Oxford University Press for International Institute for Strategic Studies, 1968).

23. Hedley Bull, "The New Balance of Power in Asia and the Pacific," *Foreign Affairs* 49, no. 4 (July 1971): 669–81.

24. Paul Dibb, *Towards a New Balance of Power in Asia*, Adelphi Paper 295 (Oxford: Oxford University Press for International Institute for Strategic Studies, 1995).

25. Jan Hornet, "Say No to a Balance of Power in Asia," *National Interest*, March 31, 2014, http://nationalinterest.org/commentary/say-no-balance-power-asia-10159.

26. David C. Kang, *China Rising: Peace, Power, and Order in East Asia* (New York: Columbia University Press, 2007).

27. Alastair I. Johnston, "What (If Anything) Does East Asia Tell Us about International Relations Theory?" *Annual Review of Political Science* 15 (2012): 59.

28. Steve Chan, *Looking for Balance: China, the United States, and Power Balancing in East Asia* (Stanford, CA: Stanford University Press, 2012), 4.

29. For further reading on the concept of "hedging," see Evelyn Goh, "Understanding 'Hedging' in Asia-Pacific Security," *PacNet* 43 (August 31, 2006).

30. For further reading on this possibility, see Nicholas Khoo, "Is Realism Dead? Academic Myths and Asia's International Politics," *Orbis* 58, issue 2 (2014): 182–97.

31. See Amitav Acharya, *Constructing a Security Order in Southeast Asia: ASEAN and the Problem of Regional Order* (New York: Routledge, 2009).

32. Evelyn Goh, "Great Powers and Hierarchical Order in Southeast Asia: Analyzing Regional Security Strategies," *International Security* 32, issue 3 (Winter 2007–8): 113–57.

33. Hillary Clinton, "America's Pacific Century," *Foreign Policy* 189 (November 2011): 56–63.

34. It should be noted that the United States does have "observer" status in CICA.

35. David Shambaugh, "Introduction: The Rise of China and Asia's New Dynamics," in *Power Shift: China and Asia's New Dynamics*, ed. David Shambaugh (Berkeley: University of California Press, 2005), 16.

36. Amitav Acharya, "Power Shift or Paradigm Shift? China's Rise and Asia's Emerging Security Order," *International Studies Quarterly* 58 (2014): 158–73.

Glossary

absolute monarchy: A monarchical form of government in which the monarch has absolute power among his or her people.

anarchic regime: A type of international organization in which each member state retains veto rights.

anarchy: A system in which there is no global law enforcement authority to manage international conflict, enforce agreements, or guarantee the survival of states.

anti-access / area denial (A2/AD): Weaponry capable of deterring enemy forces from entering a strategically vital area, such as the Taiwan Strait or the South China Sea.

arms race: A pattern of military spending where two or more countries purchase and develop military equipment directly in response to the purchases and developments of opposing states. This situation is seen as harmful to regional or global stability because the action/reaction dynamic undermines trust and increases tension.

ASEAN Charter: A binding document signed in 2007 by the ten members of ASEAN. The Charter finalizes a process of reform and commits ASEAN to developing cooperation in a wide range of security, political, economic, and social areas.

ASEAN Defense Ministers' Meeting Plus (ADMM+): A meeting of the defense ministers of the ten ASEAN member states and the eight "plus" countries of the East Asia Summit—Australia, China, India, Japan, New Zealand, South Korea, Russia, and the United States.

ASEAN Regional Forum (ARF): An intergovernmental organization that draws together twenty-seven countries that have a bearing on the security of the Asia-Pacific. It was formed in 1994 to develop cooperative and preventive diplomacy across the region.

Asia-Pacific Economic Cooperation (APEC): Formed in 1989, this forum includes twenty-one members and aims to facilitate economic growth and prosperity in the Asia-Pacific via trade and investment liberalization, business facilitation, and economic and technical cooperation.

Association of Southeast Asian Nations (ASEAN): Formed in 1967, it now includes ten members and is the most successful example of multilateralism in the Asia-Pacific. Its recent Charter has cemented a period of substantial reform that now sees ASEAN members discussing domestic, economic, political, and security issues. ASEAN is the driver of current Asia-Pacific multilateral diplomacy.

authoritarian: A system of government involving tightly controlled authority that denies the public participation in choosing its leadership or participating in the development of laws. Authoritarian states are often characterized by repression against those who challenge the state, by emotional or nationalist movements that help support the leadership, and by unclear legal constraints on those in authority.

autonomy (political): Self-government; what many separatist movements seek.

balance of power: A realist idea that two or more great powers with similar capabilities can balance each other in a "bipolar" or "multipolar" order and thereby prevent conflict.

Chinese Communist Party (CCP): The sole political party that leads the post-revolution People's Republic of China. The CCP was originally a revolutionary party—viscerally opposed to all entrenched authority, identifying strongly with the oppressed and exploited classes, and focused on international revolution. It has since shifted ideologically while staying in a position of authority in Chinese politics.

coalition: a grouping of states established to address a specific issue. May evolve into more formal multilateral groupings or dissolve once the issue is resolved.

collective defense: The right to help defend others if they are under attack.

colonialism: The control or governing influence of a nation over a dependent country, territory, or people as a colony, territory, or protectorate.

Commonwealth Secretariat: The central institution responsible for facilitating cooperation and development among the fifty-three members of the Commonwealth of Nations.

communal violence: Violence between communities of different ethnic, religious, or cultural origin.

communism: A theory or system of social organization based on the holding of all property in common; actual ownership thus is ascribed to the community as a whole or to the state.

comprehensive security: A definition of security that extends beyond traditional military security to include economic, diplomatic, and human security. In addition: A security concept developed by Japan in the 1980s that extends beyond traditional military security to include food, energy, environmental, and social security vulnerabilities essential to maintaining regional security and domestic/government stability.

concert of powers: A realist idea that several great powers can jointly manage international affairs and accommodate their competing interests on the basis of common goals, values, and interests.

confidence building: A first step in establishing effective multilateralism. Confidence building is about developing a belief between two or more states that other states possess honest intentions and can be expected to honor their agreements.

Congress of Vienna: A meeting of the European powers after the Napoleonic Wars that reorganized Europe.

constructivism: A security studies theory that holds that ideational factors are as important as material power in determining the security of states.

coordination: When two or more states, through diplomatic contact, develop a common approach to shared problems.

Copenhagen School: A security studies theory that is interested in the security not only of states but also of other "human collectivities"—national economies, ideologies, collective identities, species, and habitats.

counterinsurgency: The effort of a state and its supporters and allies to defeat an insurgency.

crimes against humanity: Serious acts (e.g., rape, slavery, murder, or torture) conducted as part of a widespread or systematic attack directed against any civilian population.

criminal terrorism: Terrorism committed by groups motivated by sheer profit or a combination of profit and political motives.

critical security studies: A security studies theory that holds that all knowledge is socially constructed, and therefore opens up possibilities for reconstructing perceptions of the international system.

Cultural Revolution: A political program in China from 1966 to 1976 to eliminate noncommunist elements from society and ensure the political dominance of Mao Zedong. The program involved the purging of many senior officials and violent struggles across China.

cyberattack: Offensive actions that target computer information systems and networks.

cyber defense: Protective actions taken in anticipation of an attack against computer information systems and networks.

cyber deterrence: Actions taken to dissuade adversaries from engaging in a cyberattack.

cyberpower: The ability to use cyberspace to create advantages and influence events in other operational environments and across the instruments of power.

cyber security: Measures relating to the confidentiality, availability, and integrity of information that is processed, stored, and communicated by electronic or similar means.

cyberspace: The environment in which communication occurs over computer networks.

decolonization: The process by which states became independent of their colonizing power.

democratic/democracy: A system of government based upon popular authorization of authority. This ideally involves equal say by all eligible citizens in the proposal, development, and creation of laws. Typically, given the complexity of modern states, citizens elect representatives to make laws in their name, although citizens can sometimes vote directly on laws (especially controversial issues).

democratic peace thesis: A liberal idea based on the observation that democracies do not go to war against each other.

diplomacy: The process by which states come together to peacefully discuss their mutual interests and concerns.

dissident terrorism: Terrorism committed by nonstate groups against governments and other perceived enemies.

doctrine: A set of standards and practices to enable militaries to undertake campaigns. Doctrine helps guide militaries to act in reliable ways to achieve their objectives.

East Asia Summit (EAS): Launched in 2004, the EAS seeks to address security issues in the Asia-Pacific by bringing heads of state/government together. It has eighteen members, including the United States, China, Russia, and many of the middle powers. It is seen as the premier multilateral security institution in Asia today.

energy security: Access to sufficient energy supplies at affordable prices.

ethnic: A population subgroup (within a larger or dominant national or cultural group) with a common cultural tradition, religion, language, or other distinguishing characteristic.

ethnic cleansing: The forced removal/displacement or killing of members of one ethnic or religious group in an area by members of another group.

ethnonationalism: A strain of nationalism that is marked by the desire of an ethnic community to have absolute authority over its own political, economic, and social affairs.

Exclusive Economic Zone (EEZ): The maritime jurisdiction within which coastal states have the right to exploit, develop, manage, and conserve all resources found in the waters, on the ocean floor, and in the subsoil of an area extending 200 miles from their shores.

failed state: A state that is failing to perform the functions typically expected of states. That is, it has lost physical control of its territory or a monopoly on the legitimate use of force, it no longer has legitimate authority to make collective decisions, and it is unable to provide reasonable public services or to interact with other states as a full member of the international community.

genocide: Acts committed with the intent to destroy, in whole or in part, a national, ethnic, racial, or religious group.

globalization: The process of international integration arising from the interchange and movement of technology, people, products, ideas, and culture.

Great Leap Forward: A catastrophic policy of social reengineering in China from 1958 to 1960 that led to massive starvation, setting back the country's development significantly.

gross domestic product (GDP): The total value of all goods and services produced within a country each year.

guerrilla: Small, independent armed bands that engage in irregular fighting. Guerrilla tactics tend to emphasize troops dispersed into small groups to strike at an enemy's weak points, withdrawing before larger formations of regular troops can be mustered.

hacktivist: A person who uses computer information technology and networks to promote political ends.

Hallstein Doctrine: The doctrine that a state may recognize either North Korea or South Korea but not both.

hegemony: A realist idea according to which a single state maintains order with superior economic and military resources, dominates the system, defines collective goals and rules, and enforces them.

heterogeneous: A society that consists of different ethnic, cultural, and/or religious groups.

human security: A security studies theory that argues that the primary referent of security is the individual human person and that security policy should seek to protect populations from conflict-related violence and to advance basic human needs.

humanitarian intervention: The threat, or use, of force by a state or group of states primarily for the purpose of protecting citizens in another state from widespread crimes against humanity, ethnic cleansing, war crimes, or other serious threats to human security.

idealist: An optimistic approach to international relations that focuses on values such as encouraging cooperation, tolerance, and peace. Sometimes this term is used to dismiss the practicality of those who advocate for alternate, principled policies.

information warfare: The use of information and communication technology to pursue a competitive advantage over an opponent.

information warriors: People engaged in cybersecurity, cyber defense, cyberattack, and cyber deterrence.

institutions: Arrangements or organizations that have rules and practices that define the expectations and actions of their members.

insurgency: A protracted violent conflict in which one or more groups seek to overthrow or fundamentally change the political or social order in a state or region through the use of sustained violence, subversion, social disruption, and political action.

interdependence: A situation in which actors are affected by the decisions made by others.

international governmental organization (IGO): An association of states to regulate mutual economic and security concerns.

intervention: The threat or use of force by a state, group of states, or international organization against another state.

jihad, jihadist: In Arabic, "jihad" translates most literally as "struggle" and is a duty of devotees of Islam. The term has been co-opted, however, to refer to participating in violent struggle variously against outsiders, infidels, and other opponents of the faith. "Jihadist" is usually used as shorthand for religiously motivated fighters, usually foreign fighters.

jurisdiction: In a maritime context, this is authority over maritime space that stems from a state's sovereignty over its land. It becomes progressively more diluted the farther one goes offshore.

liberalism: A security studies theory that is optimistic about the prospects for cooperation among international actors and the prospects for a peaceful world, based on interdependence, institutions, and the democratic peace thesis.

maritime boundary disputes: Disputes over the delimitation of boundaries at sea.

Melanesian Spearhead Group: An intergovernmental organization composed of the four Melanesian states of Fiji, Papua New Guinea, the Solomon Islands, and Vanuatu—as well as the Front de Libération Nationale Kanak et Socialiste, representing the Kanak population of New Caledonia—that is intended to promote cooperation and development in the Melanesian subregion of the South Pacific.

microstate: A state with a population of less than either 1.5 million or 100,000.

middle powers: States that can protect their core interests and initiate or lead a change in a specific aspect of the existing international order.

Millennium Development Goals: The eight international development goals that were established following the Millennium Summit of the United Nations in 2000.

minilateral: A multilateral forum that seeks to use the minimum number of member states to achieve its objectives. Tends toward ad hoc meetings and to be focused on specific issue areas.

modernization: Economic meaning: A model of a progressive economic and political transition from a "premodern" or "traditional" to a "modern" society. Military meaning: The renewal and development of a military's capabilities. This may include renewal of aging equipment, or the development and purchase of new equipment.

multilateral: A forum involving three or more states. May formalize longstanding institutions or ad hoc meetings on specific issues.

multilateralism: Diplomacy where three or more states meet together. Can take the form of one-time meetings, a series of meetings. or the creation of organizations to facilitate long-term cooperation.

multipolar: A security order in which more than two states have equal, or close to equal, power.

national identity gap: An implicit contradiction between the national identity or national role expectations of one state and another.

national interests: The key concerns of nations, typically understood in terms of avoiding armed attack, securing resources that increase prosperity, and encouraging a regional order conducive to their security and influence.

nationalism: A political ideology that involves individuals sharing a sense of national identity that generates a sense of loyalty to their nation. As a motive for terrorism or insurgency, a desire to be free of the influence or control of another state or nation.

neoliberalism: A variation of liberalism. This security studies theory argues that the negative effects of anarchy can be mitigated by states cooperating to form international institutions.

neorealism: A variation of realism. This security studies theory argues that the most important factor in shaping state behavior is the structure of the international system. This is measured by understanding the wealth and capacity of the major states and their alliance relationships.

nonalignment: An attempt to maintain good relations with the superpowers, the West, and the developing world, without making alliances.

nontraditional security: Security concept referring to nonmilitary, transnational threats to peoples and states that require comprehensive and multilateral remedies, such as transnational organized crime, infectious diseases, drug trafficking, human smuggling, environmental degradation, and natural disasters.

norm: Shared expectations about appropriate behavior.

normal nation: A nation that assumes international responsibilities and has sufficient economic, diplomatic, and military capability to defend itself and its partners.

offensive realists: A variation of realism. Offensive realists are pessimistic about the chances for peace, given that they argue that the anarchy of the international system means that states can never be certain about each other's intentions and must therefore compete.

one-China principle: The diplomatic principle, often adopted by "divided nations," that any nation recognizing the rival claimant to national identity (in this case, the Republic of China or Taiwan) cannot then have diplomatic relations with the People's Republic of China in Beijing—in other words, "one (and only one) China."

Pacific Community: An international organization that attempts to promote sustainable development in the South Pacific region.

Pacific Islands Forum: An intergovernmental organization that aims to enhance cooperation between independent countries in the South Pacific region.

pacifism: The rejection of war as a sovereign right of states.

piracy: According to the International Maritime Bureau, this is the act of boarding or attempting to board any ship with the apparent intent to commit theft or any other crime and with the apparent intent or capability to use force in the furtherance of this act. When this occurs on the high seas, it is called piracy. When it occurs inside a state's territorial waters, it is called armed robbery at sea.

pluralism: Where smaller groups within society maintain their own identity, such as ethnic, linguistic, or religious identity.

power: In simple terms, the ability of one actor to make another actor do what it would otherwise not do. This can also include the ability to influence other actors to change their beliefs.

power transition: The theory that when a rising power overtakes and surpasses an established power, it is a very sensitive time for both, and a time when a major war is likely to occur.

predictability: A quality that defines how a state views the actions and activities of another state. Greater levels of predictability ensure that states have more accurate understandings of the motives and interests of other states, which helps make their own policies more effective.

preventive diplomacy: A sophisticated version of multilateralism, preventive diplomacy is where states proactively discuss issues in order to stop them from becoming a threat to peaceful relations later on. This often involves a willingness to discuss sensitive domestic issues.

Proliferation Security Initiative: A multistate effort to build capabilities to counter illicit trade in technology for weapons of mass destruction, their delivery systems and related materials, to and from states and nonstate actors of proliferation concern.

realism: A security studies theory that is pessimistic about the prospects of the security of states, which it holds act according to self-interested impulses in order to ensure their security. Differing approaches to realism propose alternative methods whereby states can ensure their survival.

referent object: The object whose security is being analyzed.

regionalism: An attempt to provide a coherent sense of identity and purpose to a geographic area, often through multilateral institutions. Regionalism argues that there are important and defining links within large territorial areas that may have a greater impact on international relations than the borders of states or their ideological or ethnic organization.

religious terrorism: Terrorism committed by groups for the greater glory of their faith.

responsibility to protect: A principle adopted at the 2005 UN World Summit that holds that (1) each state bears a primary responsibility to protect its population from genocide, war crimes, crimes against humanity, and ethnic cleansing; (2) the international community has a responsibility to assist states to uphold this responsibility; and (3) the international community bears a responsibility to take timely and decisive action, including force as a last resort, if a state manifestly fails to protect its populations from the four atrocity crimes.

sea lanes of communication (SLOC): Designated passages at sea used for commerce.

secessionist movement/secessionism: A movement by a national, ethnic, or other community to break away from the sovereign state in which they reside in order to create a new sovereign state.

securitization: The idea that security is a "speech act"—that is, what defines what is and is not a security issue is whether people describe it as a security issue, and if those who listen accept this description.

security community: An idea first put forward by Karl Deutsch in the 1950s to describe an area of the world where states no longer consider violence or war a suitable policy option in their mutual relations. Political elites are bound together by shared understandings, and the citizens of states are similarly closely tied together. Constructivists in the 1990s emphasized the valuable role that shared identity plays in these communities.

security dilemma: A situation in which a state's desire for security forces it to acquire more power, which in turn makes other states insecure and motivates them to increase their own power.

security order: A situation that exists when interaction among states is not arbitrary but conducted in a systematic manner on the basis of certain rules.

self-help: A situation in which states must take care of their own security by strengthening their military power and forming tactical, but temporary, alliances with other states to balance against threats.

separatism: One group's desire to be separated from a larger group, based on identity differences, such as ethnicity, language, or religion.

small state: A state that has fewer economic, military, and societal resources than great and middle powers and that consequently is vulnerable to interference from larger powers and finds it difficult to act independently.

soft authoritarian: A political system in which a category of political parties or organizations has successively won election victories and whose future defeat cannot be envisaged or is unlikely for the foreseeable future.

soft power: A concept invented by the Harvard University professor Joseph S. Nye Jr. in the 1990s to describe the capacity of states to attract support, rather than forcing or coercing it.

sovereignty: A state's absolute control over the land within its territorial borders. States cannot have sovereignty over ocean space.

state building: A process whereby external actors attempt to help a state achieve control over its territory, gain the loyalty of its population, and build durable, centralized institutions that hold a monopoly over violence.

state terrorism: Terrorism committed by governments, either internationally or domestically, against perceived enemies.

strategic condominium: See "concert of powers."

subsistence farmers: Farmers who focus on growing enough food to feed themselves and their family, leaving little or no surplus to be marketed.

Sustainable Development Goals: The seventeen international development goals that were adopted in September 2015 to build on and expand the UN Millennium Development Goals.

terra nullius: Originally, a Roman legal term; now embedded in international law to mean land that belongs to no one. Often applied by states seeking to make claims to new territory and seeking to deny that anyone else has prior claims or authority.

territorial disputes: Disputes over the sovereignty of a rock or island.

terrorism: The deliberate creation and exploitation of fear through violence or the threat of violence in the pursuit of political change.

track-one diplomacy: Meetings of state representatives, at whatever level, in formal discussions.

track-two diplomacy: Meetings of individuals and organizations outside the formal political apparatus of the state. Track-two diplomacy is often thought of as an incubator for new and sensitive ideas that can then be discussed within track-one processes.

transnational terrorism: Terrorism that crosses national boundaries, whether in terms of support, personnel, or the location of attacks.

Trans-Pacific Partnership: A United States–led trade agreement negotiated between twelve Pacific Rim countries—Australia, Brunei, Canada, Chile, Japan, Malaysia, Mexico, New Zealand, Peru, Singapore, the United States, and Vietnam.

trust: A quality of interstate relations that colors mutual interpretations of actions. The more trust there is, the greater the likelihood that states will understand each other as nonthreatening and that any particular action does not mean they are being targeted for aggression.

trust building: A process of diplomatic contact explicitly designed to promote trust in mutual relations. It tends to be open-ended, and it starts with creating regular contact between political elites.

unilateral: An act undertaken by a single state that is seen as having notable consequences for other states.

United Nations: An intergovernmental organization founded in 1945 in order to promote international cooperation.

United Nations Convention on the Law of the Sea (UNCLOS): This treaty codifies the law of the sea. It has 165 state parties and came into effect in 1994.

United Nations Industrial Development Organization: A United Nations institution that attempts to achieve the promotion and acceleration of industrial development.

war crime: A serious and large-scale violation of international humanitarian law committed during an armed conflict. Examples include but are not limited to deliberately targeting civilian populations or civilian infrastructure, torture, conflict-related sexual violence, and conscripting child soldiers.

Westphalian state system: The division of international society into sovereign state entities following the Treaty of Westphalia in 1648.

World Bank: A United Nations international financial institution that provides loans to developing countries for capital programs.

youth bulge: A demographic condition in which a disproportionate amount of a country's population is age fifteen to twenty-four years.

zero-sum game: The idea that all power gains are relative (i.e., for every gain one state makes, another loses something).

Contributors

Andrew Carr is a research fellow in the Strategic and Defense Studies Centre at the Australian National University. His publications include work on Australian foreign and defense policy, middle power theory, and Asian security.

Alistair D. B. Cook is a coordinator of the Humanitarian Assistance and Disaster Relief Programme and research fellow at the Centre for Non-Traditional Security Studies at the S. Rajaratnam School of International Studies at Nanyang Technological University in Singapore. In 2012–13 he was a visiting research fellow at the East Asian Institute at the National University of Singapore.

Mathew Davies is a senior lecturer in the Department of International Relations at the Australian National University. His research examines the intersection of regional order building, human rights, and governance in Southeast Asia, paying particular regard to the Association of Southeast Asian Nations.

Lowell Dittmer is a professor in the Department of Political Science at the University of California, Berkeley. His current research interests include a study of China–Asia relations and an analysis of current Chinese political morality.

H. D. P. (David) Envall is a research fellow and undergraduate convenor in the Department of International Relations at the Australian National University. His research focuses on Japanese security and foreign policy, political leadership, and the international relations in the Asia-Pacific region.

Brad Glosserman is executive director of the Pacific Forum CSIS (Center for Strategic and International Studies) in Honolulu, which has provided policy-oriented analysis and promoted dialogue on regional security, political, economic, and environmental issues in the Asia-Pacific region for more than twenty-five years. He oversees all Pacific Forum programs, conferences, and publications. He is the author, with Scott Snyder, of *The Japan–South Korea Identity Crisis: East Asian Security and the United States* (Columbia University Press, 2015).

Ian Hall is a professor in the School of Government and International Relations at Griffith University. His research focuses on the intellectual history of international relations and Indian foreign policy.

Rex B. Hughes is a codirector of the Cyber Innovation Network in the Computer Laboratory at the University of Cambridge, and a visiting fellow for cybersecurity at Wolfson College at the University of Cambridge and at the Munk School of Global Affairs at the University of Toronto.

Tim Huxley is executive director of the International Institute for Strategic Studies–Asia in Singapore. He has worked for many years at the overlap between strategic studies and Asian area studies; his research focuses particularly on Southeast Asian states' security and defense policies.

James Manicom is the author of *Bridging Troubled Waters: China, Japan, and Maritime Order in the East China Sea* (Georgetown University Press, 2014) and an expert on East Asia, the Arctic, and global security. He received a BA in international relations from Mount Allison University and an MA and PhD in international relations from Flinders University in Australia.

Nick Nelson is a staff member of the Centre for Defense and Security Studies at Massey University. His current research focuses on Southeast Asia and Pacific security, terrorism and political violence, and leadership and management.

Christopher Paul is a senior social scientist at the RAND Corporation. His current research interests include counterinsurgency, irregular warfare, security cooperation, and information operations.

Brendan Taylor is head of the Strategic and Defense Studies Centre at the Australian National University. His publications include work on Korean Peninsula security issues, US-China relations, economic sanctions, and the Asia-Pacific region's security architecture.

Sarah Teitt is deputy director and researcher at the Asia-Pacific Centre for the Responsibility to Protect of the University of Queensland, where she is responsible for advancing research and building partnerships aimed at the prevention of genocide and other mass atrocities in the Asia-Pacific region.

William T. Tow is a professor of international relations and head of the Department of International Relations at the Australian National University. He is codirector of the Australian National University's projects for the MacArthur

Foundation's Asia Security Initiative and for the Centre of Excellence in Policing and Security.

Joanne Wallis is a senior lecturer in the Strategic and Defense Studies Centre at the Australian National University. From 2012 to 2014, she was convenor of the Centre's Asia-Pacific Security program. Her publications include works on state building, constitution making, nation building, and peace building in the South Pacific.

Index

Figures, maps, notes, and tables are indicated by f, m, n, and t following the page number.

CPSIA information can be obtained
at www.ICGtesting.com
Printed in the USA
BVHW07s0235190918
527875BV00019B/186/P